Lecture Notes in Artificial Intelligence 11054

Subseries of Lecture Notes in Computer Science

More information about this series at http://www.springer.com/series/1244

Carlos Alzate · Anna Monreale et al. (Eds.)

ECML PKDD 2018 Workshops

MIDAS 2018 and PAP 2018
Dublin, Ireland, September 10–14, 2018
Proceedings

 Springer

Editors
Carlos Alzate 🆔
IBM Research - Ireland
Dublin, Ireland

Anna Monreale 🆔
KDD Lab
University of Pisa
Pisa, Italy

Workshop Editors *see next page*

ISSN 0302-9743   ISSN 1611-3349 (electronic)
Lecture Notes in Artificial Intelligence
ISBN 978-3-030-13462-4   ISBN 978-3-030-13463-1   (eBook)
https://doi.org/10.1007/978-3-030-13463-1

Library of Congress Control Number: 2019931847

LNCS Sublibrary: SL7 – Artificial Intelligence

This Springer imprint is published by the registered company Springer Nature Switzerland AG
The registered company address is: Gewerbestrasse 11, 6330 Cham, Switzerland

Workshop Editors

Livio Bioglio (iD)
University of Turin
Turin, Italy

Valerio Bitetta
R&D Department
UniCredit
Milan, Italy

Ilaria Bordino
R&D Department
UniCredit
Rome, Italy

Guido Caldarelli
IMT School for Advanced Studies Lucca
Lucca, Italy

Andrea Ferretti
R&D Department
UniCredit
Milan, Italy

Riccardo Guidotti (iD)
KDD Lab
University of Pisa
Pisa, Italy

Francesco Gullo
R&D Department
UniCredit
Rome, Italy

Stefano Pascolutti
R&D Department
UniCredit
Milan, Italy

Ruggero G. Pensa (iD)
Department of Computer Science
University of Turin
Turin, Italy

Celine Robardet (iD)
National Institute of Applied Science
Université de Lyon
Lyon, France

Tiziano Squartini
IMT School Advanced Studies Lucca
Lucca, Italy

Preface

The European Conference on Machine Learning and Principles and Practice of Knowledge Discovery in Databases (ECML PKDD) is the premier European machine learning and data mining conference and builds upon over 16 years of successful events and conferences held across Europe. This year the conference — ECML PKDD 2018 — was held in Dublin, Ireland, during September 10–14, 2018. It was complemented by a workshop program, where each workshop is dedicated to specialized topics, to cross-cutting issues, and to upcoming trends. This year, 19 workshop proposals were submitted, and after a careful review process, which was led by the workshop co-chairs, 17 workshops were accepted. The workshop program included the following workshops:

1. The Third Workshop on Mining Data for Financial Applications (MIDAS)
2. The Second International Workshop on Personal Analytics and Privacy (PAP)
3. New Frontiers in Mining Complex Patterns
4. Data Analytics for Renewable Energy Integration (DARE)
5. Interactive Adaptive Learning
6. The Second International Workshop on Knowledge Discovery from Mobility and Transportation Systems (KnowMe)
7. Learning with Imbalanced Domains: Theory and Applications
8. IoT Large-Scale Machine Learning from Data Streams
9. Artificial Intelligence in Security
10. Data Science for Human Capital Management
11. Advanced Analytics and Learning on Temporal Data
12. The Third Workshop on Data Science for Social Good (SoGood)
13. Urban Reasoning from Complex Challenges in Cities
14. Green Data Mining, International Workshop on Energy-Efficient Data Mining and Knowledge Discovery
15. Decentralized Machine Learning on the Edge
16. Nemesis 2018: Recent Advances in Adversarial Machine Learning
17. Machine Learning and Data Mining for Sports Analytics (MLSA)

Each workshop had an independent Program Committee, which was in charge of selecting the papers. The success of the ECML PKDD 2018 workshops depends on the work of many individuals. We thank all workshop organizers and reviewers for the time and effort invested. We would also like to express our gratitude to the members of the Organizing Committee and the local staff who helped us. Sincere thanks are due to Springer for their help in publishing the proceedings.

This volume includes the selected papers of the MIDAS and PAP workshops. The papers of the other workshops will be published in separate volumes. Lastly, we thank all participants and keynote speakers of the ECML PKDD 2018 workshops for their contributions that made the meeting really interesting.

October 2018   Carlos Alzate
                                                               Anna Monreale

Contents

PAP 2018: The 2nd International Workshop on Personal Analytics and Privacy

MIDAS 2018: The 3rd Workshop on MIning DAta for Financial ApplicationS

MIDAS 2018: The 3rd Workshop on MIning DAta for Financial ApplicationS

Workshop Description

Motivation. Like the famous King Midas, popularly remembered in Greek mythology for his ability to turn everything he touched with his hand into gold, the wealth of data generated by modern technologies, with widespread presence of computers, users and media connected by Internet, is a goldmine for tackling a variety of problems in the financial domain.

Nowadays, people's interactions with technological systems provide us with gargantuan amounts of data documenting collective behavior in a previously unimaginable fashion. Recent research has shown that by properly modeling and analyzing these massive datasets, for instance representing them as network structures, it is possible to gain useful insights into the evolution of the systems considered (i.e., trading, disease spreading, political elections). Investigating the impact of data arising from today's application domains on financial decisions is of paramount importance. Knowledge extracted from data can help gather critical information for trading decisions, reveal early signs of impactful events (such as stock market moves), or anticipate catastrophic events (e.g., financial crises) that result from a combination of actions, and affect humans worldwide.

The importance of data-mining tasks in the financial domain has been long recognized. For example, in the Web context, changes in the frequency with which users browse news or look for certain terms on search engines have been correlated with product trends, level of activity in certain given industries, unemployment rates, or car and home sales, as well as stock-market trade volumes and price movements. Other core applications include forecasting the stock market, predicting bank bankruptcies, understanding and managing financial risk, trading futures, credit rating, loan management, bank customer profiling. Despite its well-recognized relevance and some recent related efforts, data mining in finance is still not stably part of the main stream of data-mining conferences. This makes the topic particularly appealing for a workshop proposal, whose small, interactive, and possibly interdisciplinary context provides a unique opportunity to advance research in a stimulating but still quite unexplored field.

Objectives and Topics. The aim of the *3rd Workshop on MIning DAta for financial applicationS* (MIDAS 2018), held in conjunction with the *2018 European Conference on Machine Learning and Principles and Practice of Knowledge Discovery in Databases* (ECML-PKDD 2018), Dublin, Ireland, September 10–14, 2018, was to discuss challenges, potentialities, and applications of leveraging data-mining tasks to tackle problems in the financial domain. The workshop provided a premier forum for sharing findings, knowledge, insights, experience and lessons learned from mining data generated in various domains. The intrinsic interdisciplinary nature of the workshop promoted the interaction between computer scientists, physicists, mathematicians, economists and financial analysts, thus paving the way for an exciting and stimulating environment involving researchers and practitioners from different areas.

Topics of interest included: forecasting the stock market, trading models, discovering market trends, predictive analytics for financial services, network analytics in finance, planning investment strategies, portfolio management, understanding and managing financial risk, customer/investor profiling, identifying expert investors, financial modeling, measures of success in forecasting, anomaly detection in financial data, fraud detection, discovering patterns and correlations in financial data, text mining and NLP for financial applications, financial network analysis, time series analysis, pitfall identification.

Outcomes. MIDAS 2018 was structured as a *half-day* workshop. We encouraged submissions of regular papers (long or short), and extended abstracts. Regular papers could be up to 15 pages (long papers) or 8 pages (short papers), and reported on novel, unpublished work that might not be mature enough for a conference or journal submission. Extended abstracts could be up to 5 pages long, and presented work-in-progress, recently published work fitting the workshop topics, or position papers. All submitted papers were peer-reviewed by three reviewers from the program committee, and selected on the basis of these reviews. MIDAS 2018 received 11 submissions, among which 8 papers were accepted (6 long regular papers, 1 short regular paper, and 1 extended abstract). In accordance with the reviewers' scores and comments, the paper entitled "*A progressive resampling algorithm for finding very sparse investment portfolios*", authored by Marko Hassinen and Antti Ukkonen, was selected as the best paper of the workshop.

Organization

Program Chairs

Valerio Bitetta	UniCredit, Italy
Ilaria Bordino	UniCredit, Italy
Guido Caldarelli	IMT Lucca, Italy
Andrea Ferretti	UniCredit, Italy
Francesco Gullo	UniCredit, Italy
Stefano Pascolutti	UniCredit, Italy
Tiziano Squartini	IMT Lucca, Italy

Program Committee

Aris Anagnostopoulos	Sapienza University, Italy
Ioannis Anagnostou	ING Financial Markets, Netherlands
Argimiro Arratia	Universitat Politécnica de Catalunya, Spain
Antonia Azzini	C2T, Italy
Xiao Bai	Yahoo Research, USA
Paolo Barucca	University of Zurich, Switzerland
Ludovico Boratto	Eurecat, Spain
Alejandra Cabaña	Universitat Autónoma de Barcelona, Spain
Olivier Caelen	Atos Wordline, Belgium
Annalina Caputo	Trinity College Dublin, Ireland
Carlotta Domeniconi	George Mason University, USA
Jerome François	Inria, France
João Gama	University of Porto, Portugal
Roberto Interdonato	CIRAD, France
Alexandros Iosifidis	Aarhus University, Denmark
Andreas Kaltenbrunner	NTENT, Spain
Dragi Kocev	Jožef Stefan Institute, Slovenia
Nicolas Kourtellis	Telefonica Research, Spain
Iordanis Koutsopoulos	Athens University of Economics and Business, Greece
Elisa Letizia	European Central Bank, Germany
Matteo Manca	Zurich Insurance, Switzerland
Stefania Marrara	C2T, Italy
Yelena Mejova	Qatar Computing Research Institute, Qatar
Iris Miliaraki	Schibsted Media Group, Spain
Davide Mottin	Hasso Plattner Institute, Germany

A Multivariate and Multi-step Ahead Machine Learning Approach to Traditional and Cryptocurrencies Volatility Forecasting

Jacopo De Stefani[1]([✉]) [iD], Olivier Caelen[2] [iD], Dalila Hattab[3] [iD],
Yann-Aël Le Borgne[1] [iD], and Gianluca Bontempi[1] [iD]

[1] MLG, Departement d'Informatique, Université Libre de Bruxelles,
Boulevard du Triomphe CP212, 1050 Brussels, Belgium
{jacopo.de.stefani,yleborgn,gianluca.bontempi}@ulb.ac.be
[2] Worldline SA/NV R&D, Brussels, Belgium
olivier.caelen@worldline.com
[3] Equens Worldline R&D, Lille, Seclin, France
dalila.hattab@equensworldline.com

Abstract. Multivariate time series forecasting involves the learning of historical multivariate information in order to predict the future values of several quantities of interests, accounting for interdependencies among them. In finance, several of this quantities of interests (stock valuations, return, volatility) have been shown to be mutually influencing each other, making the prediction of such quantities a difficult task, especially while dealing with an high number of variables and multiple horizons in the future. Here we propose a machine learning based framework, the DFML, based on the Dynamic Factor Model, to first perform a dimensionality reduction and then perform a multiple step ahead forecasting of a reduced number of components. Finally, the components are transformed again into an high dimensional space, providing the desired forecast. Our results, comparing the DFML with several state of the art techniques from different domanins (PLS, RNN, LSTM, DFM), on both traditional stock markets and cryptocurrencies market and for different families of volatility proxies show that the DFML outperforms the concurrent methods, especially for longer horizons. We conclude by explaining how we wish to further improve the performances of the framework, both in terms of accuracy and computational efficiency.

G. Bontempi acknowledges the support of the INNOVIRIS SecurIT project *BruFence: Scalable machine learning for automating defense system*. J. De Stefani acknowledges the support of the ULB-Worldline Collaboration Agreement. Computational resources have been provided by the Consortium des Équipements de Calcul Intensif (CÉCI), funded by the Fonds de la Recherche Scientifique de Belgique (F.R.S.-FNRS) under Grant No. 2.5020.11.

© Springer Nature Switzerland AG 2019
C. Alzate et al. (Eds.): MIDAS 2018/PAP 2018, LNAI 11054, pp. 7–22, 2019.
https://doi.org/10.1007/978-3-030-13463-1_1

Keywords: Multivariate time series forecasting ·
Volatility forecasting · Multi-step ahead forecast ·
Dynamic factor models

1 Introduction

The problem of time series forecasting, in its simplest form, deals with the prediction of a given quantity of interest in the future, given its historical values. Moreover, one could be interested in forecasting the immediate next value in the future (one-step-ahead forecasting) as well as being concerned with the estimation of a sequence of future values (multi-step-ahead forecasting). In a similar fashion, the problem might involve a single quantity (univariate forecasting), or several quantities at once (multivariate forecasting), in order to exploit potential interrelationships among them. In the context of finance, specific quantities of interest are: the stock price of a given company over time, its returns or the intensity of the fluctuations affecting the price (i.e. its volatility), among others. Specifically, in the case of stock markets, the underlying trend of the market influences all the stocks that are currently traded. As shown in [18], stock prices of firms acting on the same market often show similar patterns in the sequel of news that are important for the entire market. Moreover, analyzing global volatility transmission, Engle et al. [12] found evidence supporting volatility interdependence among the world's major trading areas. For these reasons, while modeling these time dependent quantities of interest, a multivariate model appears to be a natural choice to incorporate interdependencies into the forecasting process.

Among all the quantities of interest, in the following, we will focus on the problem of multivariate volatility forecasting. In this specific case, the quantity of interest is a latent variable, which cannot be directly observed given the time series, but only estimated, according to the granularity and the type of the available data, through different measures, named volatility proxies [27]. According to the choice of the proxy, several approaches have been proposed to tackle this multivariate problem. The largest body of the volatility forecasting literature focus on multivariate extensions of the well known GARCH [2] on traditional stock market data, for instance, citing some recently published work: [13] and [3]. For a thorough review of the different univariate and multivariate methods, we refer the interested reader to the latter. Due to the steadily growth of the cryptocurrencies market capitalization [11], coupled with the currencies' volatility, GARCH-like models [7], [32] have also been applied for non-traditional markets. The main problem of these approaches is that traditional multivariate models often suffers from the "curse of dimensionality": as the number of dimensions increase, the number of parameters grows superlinearly in the number of dimensions, making the model estimation heavily computationally intensive, especially in the case of multiple step ahead forecasts.

In order to profit from the richness of a multivariate model, while maintaining a reasonable computational complexity, we propose to employ the DFML [4], a multivariate, multistep-ahead machine learning forecasting framework involving a dimensionality compression process, based on the dynamic factor model

(DFM) principle [14]. The choice of this generic time series forecasting framework requires the usage of model-independent volatility proxies which will be discussed in Sect. 3, requiring us to dismiss GARCH as a proxy of volatility, due to his tight coupling between the proxy and the corresponding forecasting model to use, as discussed in [8].

At the time of writing, we had been able to find either multivariate techniques dealing with the forecasting of either cryptocurrencies prices [1,6] or univariate techniques dealing with the forecasting of volatility either with a one-step ahead [7,32] or multistep-ahead [10]. Nevertheless, we are not aware of any other work tackling both the problems of multivariate and multi-step ahead cryptocurrencies volatility forecasting, specifically in the case of large dimensionality and a reduced number of datapoints. Our technique will then be tested on two different benchmarks: one concerning cryptocurrencies and a second one, concerning a traditional regulated stock market (CAC40) being a de facto multivariate extension of [25].

The rest of the paper is structured as follows: Sect. 2 provides an oveview of the Dynamic Factor Machine Learner approach. Section 3 introduces the different tested multivariate models as well as the considered datasets and the formulation of the relevant forecast quantities. Section 4 concludes the paper with a discussion of the results and the future research directions.

2   Dynamic Factor Machine Learner

A Dynamic Factor Model (DFM) is a technique for multivariate forecasting originating in the economic forecasting community [14]. The basic idea of DFM is that a small number of unobserved series (or factors) can account for the temporal behavior of a much larger number of variables. If we are able to obtain accurate estimates of these factors, the forecasting endeavor can be made simpler by using the estimated dynamic factors for forecasting instead of using all series. In equations:

$$Y_{t+1} = WZ_{t+1} + \epsilon_{t+1} \tag{1}$$
$$Z_{t+1} = A_t Z_t + \cdots + A_{t-m+1} Z_{t-m+1} + \eta_{t+1} \tag{2}$$

where Y_t is a multivariate time series vector at time t, Z_t is the vector of unobserved factors of size q ($q \ll n$), A_i are $q \times q$ coefficient matrices, W is the matrix ($n \times q$) of dynamic factor loadings and the vectors of disturbances terms η are assumed to be uncorrelated. As shown in Eq. 2, in the original DFM, the latent factors follow a VAR time series process. For a detailed review of DFM models, the interested reader could refer to [28].

Here, we propose to employ a machine learning extension of the DFM (called DFML - Dynamic Factor Machine Learner). The DFML, first proposed by Bontempi et al. [4] and further discussed in [9], relies on dimensionality reduction techniques to extract the factors. Then, the factors are forecast using a nonlinear model. Finally, the forecasts of the factors are transformed back to the original values by inverting the dimensionality reduction process. The basic architecture of the DFML is depicted in Fig. 1, along with the description of the different variants. Concerning dimensionality reduction, both linear (PCA) and nonlinear (autoencoder) techniques are employed in the DFML. Linear dimensionality reduction by PCA transforms the n original time series $\mathbf{Y}[1], \ldots, \mathbf{Y}[n]$ into q new variables $\mathbf{Z}[1], \ldots, \mathbf{Z}[q]$ (called *principal components* or *factors*) such that the new variables are uncorrelated with each other and account for decreasing portions of the variance of the original variables. The q principal components are then expressed as weighted sums of the elements of $\mathbf{Y}$ with maximal variance, where the weights are normalized and constrained to ensure orthogonality:

$$\mathbf{Z}[p] = \sum_{j=1}^{n} w_{jp} \mathbf{Y}[j], \qquad p = 1, \ldots, q \tag{3}$$

Given the multivariate time series matrix $\mathbf{Y}$, $\mathbf{Z} = \mathbf{YW}$ represents the projection of the series onto the pth principal components and $\hat{\mathbf{Y}} = \mathbf{ZW}^T$ represent the reconstruction $\hat{\mathbf{Y}}$ of the values of $\mathbf{Y}$, based on the factors $\mathbf{Z}$. On the other hand, nonlinear dimensionality reduction is performed through the use of autoencoders. Autoencoders are neural networks trained to learn identity mapping from inputs to outputs [31], through a constrained architecture to enforce dimensionality reduction. As such their input and output layer have the same number of neurons n as the number of input time series but their hidden layers contain a reduced number of neurons q. Autoencoders are composed of two stacked multi-layer networks: an *encoder*:

$$\mathbf{Z}_t = f_\theta(\mathbf{Y}_t) \tag{4}$$

that transforms inputs $\mathbf{Y}_t$ into some latent (encoded) representation $\mathbf{Z}_t$, and a *decoder*:

$$\hat{\mathbf{Y}}_t = g_{\theta'}(\mathbf{Z}_t) \tag{5}$$

that reconstructs an approximation $\hat{\mathbf{Y}}_t$ of the input $\mathbf{Y}_t$ on the basis of the latent feature $\mathbf{Z}_t$ and where the mappings f_θ and $g_{\theta'}$ are non-linear. The network is usually trained using gradient descent techniques such as backpropagation, with the objective of minimizing the mean-squared error between the input $\mathbf{Y}_t$ and the output (its reconstruction $\hat{\mathbf{Y}}_t$) [31]. Concerning the forecasting part, the original DFML paper [4] proposes to forecast each factor independently (given their orthogonality) using a nonlinear model (lazy learning [5]) and a univariate multi-step-ahead forecasting strategy. In addition to the basic forecaster, the paper also proposes an optimized version (DFML'), performing a joint selection of the hyperparameters (number of factors for the dimensionality reduction, predictor and multi-step-ahead strategy for the forecaster) using out-of-sample assessment. Although we consider lazy methods for the forecaster, the modular architecture of this framework easily allows the replacement of the aforementioned technique with alternative supervised machine learning approaches (e.g. SVM, RNN).

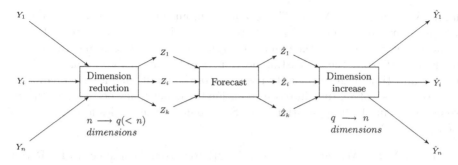

	Dimensionality reduction	Forecast
DFM	PCA	VAR
DFML$_{PCA}$	PCA	Lazy-learning
DFML$_A$	Autoencoder	Lazy-learning
DFML'$_{PCA}$	Optimized PCA	Optimized Lazy-learning

Fig. 1. Schema of the DFML architecture with a summary of the different components as implemented in the different proposed methods.

3   Methodology

3.1   Multivariate Forecasting Methods

Multiple Univariate Techniques - {Naive, UNI}: In the case of a multivariate time series **Y**, *univariate* approaches are still of interest since the multivariate forecasting task can be decomposed in a number of single-output multi-input tasks (or equivalently in a set of NARX tasks with exogenous variables)

$$
\begin{cases}
Y_{t+1}[1] &= f_1(Y_t[1], \ldots, Y_{t-m+1}[1], \ldots, \\
& \quad Y_t[n], \ldots, Y_{t-m+1}[n]) + w_t[1] \\
\vdots \\
Y_{t+1}[n] &= f_n(Y_t[1], \ldots, Y_{t-m+1}[1], \ldots, \\
& \quad Y_t[n], \ldots, Y_{t-m+1}[n]) + w_t[n]
\end{cases} \tag{6}
$$

In this case the training set is used to learn the n mapping functions f_i, $i = 1, \ldots, n$, with $w_t[i]$ being uncorrelated disturbances. For large n the problem of large input dimensionality can be addressed by adopting a feature selection technique, selecting a reduced number q of most correlated features For these univariate techniques, we will also consider a *naive* method in which $\forall i \in \{1, \ldots, n\}, f_i(t) = Y_{t-1}[i]$, i.e. for every series, the forecast for the following H steps is given by the last available value. These are the baseline methods against which we will compare the performances of our forecaster.

Partial Least Squares - PLS: Partial Least Squares [15] allows the joint forecasting of the H steps ahead of the multivariate time series on the basis of the lagged vectors $\mathbf{Y}_t, \ldots, \mathbf{Y}_{t-m}$. This is a multi-input multi-output regression task where the number of inputs amounts to nm and the number of outputs to Hn respectively, with n being the number of variables, m the embedding order of the model and H being the forecasting horizon. The benefit of PLS is that it allows at the same time a dimensionality reduction of the inputs and a joint prediction of the outputs, taking then into consideration the dependency between the future steps. An example of application of PLS in financial time series forecasting can be found in [22].

Recurrent Neural Networks - {RNN, LSTM}: Recurrent Neural networks (RNN) form a class of predictive models based on neural networks, in which recurrent connections to the network inputs allow to model dynamic temporal dependencies. In their simple form (also known as simple RNN) [17,23], the recurrent connections come from a *hidden state* H_t, which is also used for predicting future values Y_t:

$$\mathbf{H}_t = \sigma(\mathbf{W}_{HY}\mathbf{Y}_{t-1} + \mathbf{W}_{HH}\mathbf{H}_{t-1} + \mathbf{B}_H), \tag{7}$$

$$\mathbf{Y}_t = \mathbf{W}_{YH}\mathbf{H}_t + \mathbf{B}_Y \tag{8}$$

The matrices $\mathbf{W}_{HY}$, $\mathbf{W}_{HH}$, $\mathbf{W}_{YH}$, $\mathbf{B}_H$ and $\mathbf{B}_Y$ are the parameters (weights and biases) of the network, typically learned by gradient descent algorithms such as backpropagation through time [17]. A sigmoid activation function σ allows the modeling of nonlinear dependencies, while the recurrent connections allow the modeling of long-term temporal dependencies. Research on RNNs has recently been boosted by the advent of General Programming Graphic Processing Units (GPGPU), and improved design of the memory cell (*Long-Short Term Memory cells* [20]). These have allowed much more efficient RNN implementations, and effective training over multiple layers (*deep RNNs*). RNNs architectures have reached state-of-the-art performances for volatility either as part of an LSTM-GARCH hybrid model [21,33] or as standalone model [26].

3.2   Datasets Description

CAC40: The available data consists of 1645 data points of the 40 time series composing the french stock market index CAC40 from 02/01/2012 to 08/06/2018 (approximately 6 years and 5 months) in OHLC (Opening, High, Low, Closing) format.

Cryptocurrencies: The available data comes from the Kaggle dataset "Every Cryptocurrency Daily Market Price"[1] constituted of 785,024 observation of 1644 different cryptotokens from 28/04/2013 to 06/06/2018. However the number of available datapoints per cryptotoken is inversely proportional to the lifespan of the token itself. In other words, the further we go into the past, the fewer values we have for our analysis, as depicted in Fig. 2. For these reasons, we restricted our analysis to the period from 28/01/2017 to 06/06/2018 for which we have complete OHLC data for 291 tokens.

---

[1] https://www.kaggle.com/jessevent/all-crypto-currencies.

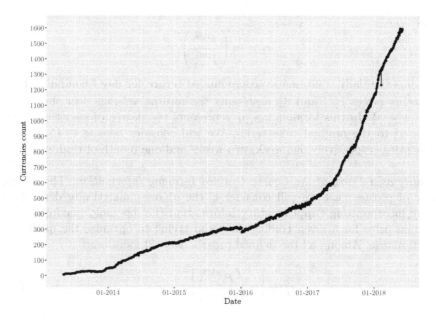

Fig. 2. Number of available datapoints for the cryptocurrencies dataset as a function of time

3.3   Volatility Proxies

The OHLC available data is composed of several quantities of interest, each of them on a daily time scale: $P_t^{(o)}, P_t^{(c)}, P_t^{(h)}, P_t^{(l)}$, respectively the stock prices at the *opening*, *closing* of the trading day and the *maximum* and *minimum* value for each trading day. In the absence of detailed information concerning the price movements within a given trading days, stock volatility becomes directly unobservable [30]. To cope with such problem, several different measures (also called proxies) have been proposed in the econometrics literature [16,19,24,27] to capture this information. However, there is no consensus in the scientific literature upon what volatility proxy should be employed for a given purpose. Nevertheless, for an empirical analysis of the use of volatility proxies in the case of univariate forecasting, the interested reader could find more details in [8].

Volatility as Variance. The first family of proxies corresponds to the natural definition of volatility [27], that is, a rolling standard deviation of a given stock's continuously compounded returns over a past time window of size n:

$$\sigma_t^{SD,w} = \sqrt{\frac{1}{w-1} \sum_{i=0}^{w-1} (r_{t-i} - \bar{r}_w)^2} \tag{9}$$

where

$$r_t = \ln\left(\frac{P_t^{(c)}}{P_{t-1}^{(c)}}\right) \tag{10}$$

represents the daily continuously compounded return for day t computed from the closing prices $P_t^{(c)}$ and $\bar{r}_w$ represents the returns' average over the period $\{t, \cdots, t-w\}$. In this formulation, w represents the degree of smoothing that is applied to the original time series. We will consider here $w \in \{5, 10, 21\}$, representing respectively one week, two weeks and one month of trading.

Volatility as a Proxy of the Coarse Grained Intraday Information. The second family of proxies that we will consider is the σ_t^i one, analytically derived by [16] by incorporating supplementary information (i.e. opening, maximum and minimum price for a given trading day) and trying to optimize the quality of the estimation. Among all the defined proxies, we will focus on:

$$\sigma_t^0 = \left[\ln\left(\frac{P_{t+1}^{(c)}}{P_t^{(c)}}\right)\right]^2 = r_t^2 \tag{11}$$

$$u = \ln\left(\frac{P_t^{(h)}}{P_t^{(o)}}\right) \qquad d = \ln\left(\frac{P_t^{(l)}}{P_t^{(o)}}\right) \qquad c = \ln\left(\frac{P_t^{(c)}}{P_t^{(o)}}\right) \tag{12}$$

where u is the normalized high price, d is the normalized low price and c is the normalized closing price.

$$\sigma_t^4 = 0.511(u-d)^2 - 0.019[c(u+d) - 2ud] - 0.383c^2 \tag{13}$$

$$\sigma_t^6 = \underbrace{\frac{a}{f} \cdot \log\left(\frac{P_{t+1}^{(o)}}{P_t^{(c)}}\right)^2}_{\text{Nightly volatility}} + \underbrace{\frac{1-a}{1-f} \cdot \hat{\sigma}_4(t)}_{\text{Intraday volatility}} \tag{14}$$

The value of $f \in [0, 1]$ represents the fraction of the trading day in which the market is closed. It is by definition bounded in the interval $[0, 1]$, In the case of CAC40, we have that $f > 1 - f$, since trading is only performed of roughly one third of the day. Here, a is a weighting parameter, whose optimal value, according to [16] is shown to be 0.17, regardless of the value of f.

After a preprocessing phase of the datasets, involving removal of missing values and proxy calculation for each time series, the data is restructured in a multivariate time series matrix form $\mathbf{Y}$ having N (number of observations) rows and n (number of variables/time series) columns. For each proxy, this matrix is such that each row $\mathbf{Y}_t$ represent a n-dimensional vector containing the value of the given proxy for of the n variables at the time t, and the scalar value $Y_t[j]$ represent the value of jth ($j = 1, \ldots, n$) variable at time t.

4   Experimental Results

The experimental study assessed and compared the methods previously discussed in the article. The methods are listed below together with the software used for the experiments. Note that, for the sake of assessment, we set the lag $m = 2$ and the maximum number of latent factors to $q = 3$ for all methods, unless specified otherwise.

1. NAIVE: univariate baseline method using the last observed value for each time series as prediction for the following H steps.
2. UNI: univariate multi-step-ahead Direct forecasting of each individual series (Eq. 6) with a feature selection process based on correlation.
3. PLS: partial-least-squares forecasting (Sect. 3.1) implemented by the function `mvr` of the R package `pls`. The optimal values for the size of the input space and the number of principal components q is determined through an out-of-sample criterion.
4. RNN: recurrent neural network implemented by the `keras_predict` function of `kerasR`[2], the R keras interface to the `keras` Deep Learning library[3] for Theano. The network is a fully-connected RNN with 10 hidden units. Since an automated setting of the number of units would not have been feasible due to an excessive computational time, this number has been set on the basis of trial and error over a small number of synthetic series.
5. LSTM: As RNN, the model is a fully connected RNN, with 10 hidden units implemented using `kerasR`. It differs from RNN as it employs LSTM cells [20] in the hidden layer, instead of regular neurons.
6. DFM: linear Dynamic Factor Model where PCA is used for factor estimation, the number of factors is set to q and the forecasting of the factors is carried out with a VAR method implemented by the `estBlackBox` function of the R package `dse`. The batch PCA is computed using the base R `eigen` function.
7. $DFML_{PCA}$: Dynamic Factor Machine Learner where PCA is used for factor estimation, the number of factors is set to q and the forecasting of each factors is carried out in a univariate manner using a local learning predictor (lazy learning [5]) and a multi-step-ahead Direct strategy.
8. $DFML_A$: it differs from $DFML_{PCA}$ because of the use of an autoencoder instead of PCA in the process of factor estimation.
9. $DFML'_{PCA}$: it differs from $DFML_{PCA}$ because of the automatic selection strategy (described in [4]): the number of factors (in the range $[1, q]$) and the multi-step-ahead strategy (among Direct, Iterated and MIMO) and the lag m are selected by an out-of-sample strategy carried out on the training set.

4.1   Results Discussion

For each multivariate dataset we performed time series cross-validation following a rolling origin strategy [29]. The size of the training set is $2N/3$ and a sequence of 50 different test sets of length H is considered.

---

[2] https://github.com/statsmaths/kerasR.
[3] https://keras.io.

For each test set, all methods are assessed in terms of the average Normalized Mean Squared Error:

$$\text{NMSE} = \frac{\sum_{j=1}^{n} \text{NMSE}[j]}{n}$$

where

$$\text{NMSE}[j] = \frac{\sum_{h=1}^{H} (Y_{T+h}[j] - \hat{Y}_{T+h}[j])^2}{V[j]H}$$

$V[j]$ is the variance of the series $Y[j]$ and $T + 1$ is the starting index of the continuation set.

While dealing with high dimensionality ($n = 291$) coupled with a relatively low number of observations ($N = 495$), as in the case of the Cryptocurrency dataset (Table 1), using the σ_t^i family of proxies, the DFML, even without hyperparameter optimisation, clearly outperforms all the concurrent methods. It should also be noted that some methods tested in the original DFML paper [4] (i.e. VAR, DSE, SSA) could not be tested due to numerical problems related to the limited number of available observations. The performances of DFML are mitigated while using proxies coming from the $\sigma_t^{SD,w}$ family, where the performance of the Naive method improves, even for forecasting horizons up to 20 steps ahead, as the smoothing provided by the window size parameter w increases. In both the cases, a linear dimensionality reduction technique with no optimization (DFM, DFML$_{PCA}$) is shown to improve the performances of the forecaster, compared to nonlinear (DFML$_A$) and optimized (DFML'$_{PCA}$) ones.

A similar ranking among the methods is observed in the case of the CAC40 dataset (Table 2), characterized by a lower dimensionality ($n = 40$) but an higher number of points ($N = 1641$). Here we can observe a generally higher average normalized NMSE, indicating a higher complexity of the forecasting problem. For the σ_t^i family, PLS and DFM appears as competitive alternatives of the DFML, especially for longer horizons ($h > 15$). As in the previous case, for the $\sigma_t^{SD,w}$ family of proxies, the performances of the DFML family are affected by the value of the smoothing factor w, where, the higher the smoothing factor is, the less effective the DFML becomes for shorter horizons, with the Naive method becoming the best one, while still maintaining good forecasting accuracy for longer horizons.

In addition to forecasting accuracy, we also analyzed the total computational time required to produce a forecast. The total computational time is obtained by summing up the time required to train the considered model and the time needed to generate a forecast. Figure 3a shows that, for low dimensionalities ($n = 40$) the total computational time of the different techniques is comparable, and independent of the forecasting horizon, except for the optimized DFML'$_{PCA}$, where the comparison of different forecasting strategies require a computational time proportional to the length of the forecasting horizon. On the other hand, for higher dimensionalities ($n = 291$), the computational time required to train multiple univariate models (UNI), neural based models (RNN and LSTM) and PLS increases considerably due to the increase of both dimensionality and forecasting

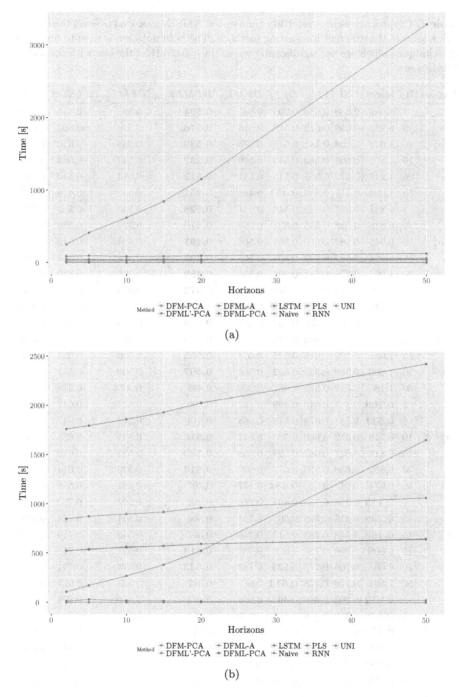

(a)

(b)

Fig. 3. Total computational time (model training + forecast) of the tested methods on the CAC40 - σ^4 (a) ($n = 40$) and cryptocurrencies - σ^4 (b) ($n = 291$) dataset-proxy combination.

Table 1. Cryptocurrencies - volatility time series: NMSE (averaged over all the continuation sets) of the different forecasting methods. The bold notation is used to highlight all techniques which are not significantly worse (pv = 0.05) than the one with the lowest NMSE score.

Dataset	H	Naive	UNI	PLS	DFM	$DFML_A$	$DFML_{PCA}$	$DFML'_{PCA}$	LSTM	RNN
σ_t^0	2	0.988	0.660	0.630	0.595	0.631	**0.594**	0.596	0.630	0.670
	5	0.982	0.646	0.613	0.579	0.613	**0.576**	0.588	0.605	0.656
	10	1.042	0.608	0.581	0.543	0.575	**0.539**	0.538	0.570	0.615
	15	1.172	0.602	0.584	0.540	0.569	**0.537**	0.547	0.563	0.599
	20	1.247	0.579	0.555	0.515	0.544	**0.512**	0.514	0.540	0.593
	50	1.024	0.517	0.503	**0.451**	0.483	**0.451**	0.466	0.479	0.521
σ_t^4	2	0.831	0.607	0.602	0.540	0.611	**0.528**	0.543	0.585	0.647
	5	0.816	0.598	0.585	0.521	0.580	**0.510**	0.522	0.559	0.638
	10	0.945	0.582	0.579	0.505	0.564	**0.491**	0.494	0.542	0.590
	15	0.924	0.582	0.592	0.508	0.565	**0.495**	0.498	0.551	0.580
	20	1.061	0.578	0.584	0.501	0.554	**0.489**	10.969	0.539	0.575
	50	0.950	0.553	0.563	0.474	0.524	**0.472**	0.476	0.510	0.543
σ_t^6	2	0.946	0.587	0.588	0.528	0.580	**0.512**	0.527	0.547	0.613
	5	1.103	0.561	0.578	0.507	0.551	**0.479**	0.480	0.531	0.587
	10	1.101	0.583	0.590	0.516	0.579	**0.499**	0.500	0.553	0.591
	15	1.041	0.592	0.616	0.525	0.574	**0.505**	0.509	0.554	0.591
	20	1.000	0.589	0.592	0.522	0.568	**0.507**	0.509	0.551	0.586
	50	1.185	0.557	0.582	0.481	0.530	0.481	**0.474**	0.514	0.560
$\sigma_t^{SD,5}$	2	**0.269**	0.351	0.648	0.499	0.662	0.500	0.524	0.647	0.739
	5	**0.511**	0.533	0.674	0.519	0.668	0.514	0.534	0.647	0.717
	10	0.719	0.612	0.669	0.523	0.647	**0.516**	0.534	0.635	0.790
	15	0.818	0.627	0.662	0.527	0.638	**0.520**	0.523	0.616	0.794
	20	0.852	0.636	0.653	0.517	0.629	**0.514**	0.526	0.614	0.771
	50	0.974	0.611	0.636	**0.484**	0.577	0.497	0.481	0.558	0.748
$\sigma_t^{SD,10}$	2	**0.113**	0.258	0.754	0.491	0.756	0.494	0.534	0.722	0.796
	5	**0.238**	0.415	0.769	0.501	0.751	0.504	0.541	0.728	0.838
	10	**0.466**	0.598	0.781	0.513	0.746	0.507	0.543	0.719	1.027
	15	0.606	0.668	0.780	0.526	0.737	**0.514**	0.558	0.713	0.898
	20	0.668	0.706	0.777	0.523	0.741	**0.522**	0.574	0.701	0.911
	50	0.891	0.726	0.778	**0.514**	0.683	0.547	0.533	0.657	0.907
$\sigma_t^{SD,21}$	2	**0.052**	0.203	0.989	0.493	0.992	0.493	0.570	0.899	1.034
	5	**0.108**	0.316	0.986	0.501	0.977	0.504	0.571	0.864	1.144
	10	**0.199**	0.522	0.987	0.513	0.958	0.514	0.573	0.891	1.065
	15	**0.295**	0.658	0.988	0.523	0.963	0.528	0.572	0.848	1.340
	20	**0.397**	0.748	0.990	0.535	0.874	0.544	0.571	0.863	1.117
	50	0.775	0.833	1.014	**0.586**	0.882	0.649	0.612	0.808	1.252

Table 2. CAC40 - volatility time series: NMSE (averaged over all the continuation sets) of the different forecasting methods. The bold notation is used to highlight all techniques which are not significantly worse (pv = 0.05) than the one with the lowest NMSE score.

Dataset	H	Naive	UNI	PLS	DFM	$DFML_A$	$DFML_{PCA}$	$DFML'_{PCA}$	LSTM	RNN
σ_t^0	2	1.332	1.047	0.972	0.969	0.987	**0.962**	1.010	1.018	1.006
	5	2.177	1.916	1.826	1.857	1.872	1.838	**1.822**	1.849	1.865
	10	1.438	1.246	**1.157**	1.173	1.184	1.164	1.155	1.164	1.184
	15	2.499	1.304	1.220	1.220	1.227	**1.209**	1.222	1.219	1.242
	20	1.566	1.227	1.155	1.153	1.163	1.174	**1.146**	1.160	1.160
	50	2.026	1.221	1.136	1.135	1.144	1.134	**1.120**	1.160	1.164
σ_t^4	2	0.585	0.504	0.463	**0.433**	0.521	0.434	0.450	0.564	0.496
	5	2.295	1.347	1.318	1.292	1.356	**1.268**	1.275	1.328	1.346
	10	1.047	1.003	0.948	0.936	0.991	**0.911**	0.946	1.014	1.018
	15	1.372	1.132	1.078	1.067	1.118	**1.048**	1.071	1.126	1.120
	20	1.272	1.023	0.948	0.926	0.977	**0.908**	0.933	1.010	1.007
	50	1.111	1.036	0.936	0.942	0.987	**0.919**	0.981	1.052	1.042
σ_t^6	2	1.780	0.854	0.805	0.776	0.859	0.767	**0.758**	0.852	0.822
	5	1.859	1.800	1.750	**1.741**	1.809	1.747	1.715	1.781	1.770
	10	1.264	1.171	1.106	1.102	1.154	**1.083**	1.118	1.149	1.139
	15	1.222	1.074	**1.001**	**0.999**	1.049	**1.001**	1.011	1.093	1.046
	20	1.332	1.185	**1.103**	1.107	1.156	1.108	1.116	1.172	1.170
	50	1.280	1.188	1.112	1.098	1.139	**1.089**	1.126	1.206	1.177
$\sigma_t^{SD,5}$	2	**0.276**	0.649	0.834	0.783	0.877	0.787	0.769	0.823	0.864
	5	**1.122**	1.275	1.304	1.289	1.355	1.242	1.215	1.329	1.352
	10	1.329	1.199	1.163	1.139	1.167	**1.095**	1.162	1.131	1.201
	15	1.408	1.149	1.095	1.068	1.113	**1.064**	1.066	1.111	1.134
	20	1.576	1.215	1.154	1.133	1.166	**1.141**	1.150	1.203	1.182
	50	2.584	1.292	1.316	1.444	**1.184**	1.243	1.192	1.229	1.273
$\sigma_t^{SD,10}$	2	**0.453**	0.667	0.901	0.805	0.964	0.805	0.788	0.827	0.881
	5	**0.698**	0.886	1.018	0.932	1.073	0.934	0.927	0.910	1.009
	10	1.133	1.010	1.044	**0.970**	1.073	1.005	1.005	1.000	1.104
	15	1.495	1.065	1.140	1.292	1.271	1.092	**1.013**	1.066	1.032
	20	1.642	1.141	1.181	1.340	1.223	1.145	**1.078**	1.108	1.178
	50	1.916	1.258	1.233	1.256	1.158	**1.144**	1.171	1.310	1.338
$\sigma_t^{SD,21}$	2	**0.033**	0.306	0.747	0.509	0.772	0.510	0.561	0.776	0.725
	5	**0.123**	0.372	0.732	0.530	0.736	0.595	0.566	0.867	0.716
	10	**0.346**	0.520	0.808	0.660	0.853	0.682	0.673	0.992	0.932
	15	**0.608**	0.680	0.862	0.771	0.893	0.795	12.315	0.970	0.868
	20	**0.827**	**0.827**	0.923	0.905	0.890	0.840	0.777	1.010	1.256
	50	1.603	1.259	1.210	1.357	1.109	**1.076**	1.311	1.282	1.585

horizons, while DFML models, thanks to the dimensionality reduction component, maintain a reduced computational time regardless of the forecasting horizon.

5 Conclusion and Future Work

The empirical analysis shows that DFML is able to produce accurate volatility forecasts, especially in the case of high-dimensional noisy series (i.e. Cryptocurrencies dataset) with non-smoothed volatility proxies σ^i, by summarizing well the intrinsic market correlations in a reduced number of factors. However, the presence of a smoothing factor (as in the $\sigma^{SD,w}$ proxies family) is shown to worsen the performances of the DFML methods. Moreover, we have shown that, thanks to the dimensionality reduction component, DFML methods can produce multi-step ahead forecasts with the same accuracy as concurrent methods with a great reduction in terms of computational cost. In order to further improve this framework we foresee different possible extensions. On one hand we believe that the use of additional volatility proxies, together with an automated variable selection process could further improve the forecasting performances. On the other hand, the use of incremental dimensionality reduction techniques could further improve the computational efficiency of the method at the expense of a small reduction in forecasting accuracy.

References

1. Alessandretti, L., ElBahrawy, A., Aiello, L.M., Baronchelli, A.: Machine learning the cryptocurrency market. arXiv preprint arXiv:1805.08550 (2018)
2. Andersen, T.G., Bollerslev, T.: ARCH and GARCH models. Encyclopedia of Statistical Sciences (1998)
3. Bollerslev, T., Patton, A.J., Quaedvlieg, R.: Multivariate leverage effects and realized semicovariance GARCH models (2018). https://doi.org/10.2139/ssrn.3164361
4. Bontempi, G., Le Borgne, Y.A., De Stefani, J.: A dynamic factor machine learning method for multi-variate and multi-step-ahead forecasting. In: 2017 IEEE International Conference on Data Science and Advanced Analytics (DSAA), pp. 222–231. IEEE (2017)
5. Bontempi, G., Taieb, S.B.: Conditionally dependent strategies for multiple-step-ahead prediction in local learning. Int. J. Forecast. **27**(3), 689–699 (2011)
6. Catania, L., Grassi, S., Ravazzolo, F.: Forecasting cryptocurrencies financial time series. In: CAMP Working Paper Series 3, BI Norwegian Business School (2018)
7. Catania, L., Grassi, S., Ravazzolo, F.: Predicting the volatility of cryptocurrency time-series. In: CAMP Working Paper Series 5, BI Norwegian Business School (2018)
8. De Stefani, J., Caelen, O., Hattab, D., Bontempi, G.: Machine learning for multi-step ahead forecasting of volatility proxies. In: 2nd Workshop on MIning DAta for Financial Applications (MIDAS). CEUR Workshop Proceedings, Aachen, vol. 1941, pp. 17–28 (2017). http://ceur-ws.org/Vol-1941/MIDAS2017_paper3.pdf

9. De Stefani, J., Le Borgne, Y.A., Caelen, O., Hattab, D., Bontempi, G.: Batch and incremental dynamic factor machine learning for multivariate and multi-step-ahead forecasting. https://doi.org/10.1007/s41060-018-0150-x

10. Degiannakis, S.: Multiple days ahead realized volatility forecasting: single, combined and average forecasts. Glob. Financ. J. **36**, 41–61 (2018)

11. ElBahrawy, A., Alessandretti, L., Kandler, A., Pastor-Satorras, R., Baronchelli, A.: Evolutionary dynamics of the cryptocurrency market. Roy. Soc. Open Sci. **4**(11), 170623 (2017)

12. Engle III, R.F., Ito, T., Lin, W.L.: Meteor showers or heat waves? Heteroskedastic intra-daily volatility in the foreign exchange market (1988)

13. Fengler, M.R., Herwartz, H., Raters, F.: Multivariate volatility models. In: Härdle, W.K., Hautsch, N., Overbeck, L. (eds.) Applied Quantitative Finance, pp. 25–37. Springer, Heidelberg (2017). https://doi.org/10.1007/978-3-540-69179-2_15

14. Forni, M., Hallin, M., Lippi, M., Reichlin, L.: The generalized dynamic factor model. J. Am. Stat. Assoc. **100**(471), 830–840 (2005). https://doi.org/10.1198/016214504000002050

15. Franses, P., Legerstee, R.: A unifying view on multi-step forecasting using an autoregression. J. Econ. Surv. **24**(3), 389–401 (2010)

16. Garman, M.B., Klass, M.J.: On the estimation of security price volatilities from historical data. J. Bus. **53**(1), 67–78 (1980)

17. Graves, A.: Supervised Sequence Labelling with Recurrent Neural Networks. Springer, Heidelberg (2012). https://doi.org/10.1007/978-3-642-24797-2

18. Hafner, C.M., Herwartz, H.: Structural analysis of portfolio risk using beta impulse response functions. Statistica Neerlandica **52**(3), 336–355 (1998)

19. Hansen, P.R., Lunde, A.: A forecast comparison of volatility models: does anything beat a garch (1, 1)? J. Appl. Econometrics **20**(7), 873–889 (2005)

20. Hochreiter, S., Schmidhuber, J.: Long short-term memory. Neural Comput. **9**(8), 1735–1780 (1997)

21. Kim, H.Y., Won, C.H.: Forecasting the volatility of stock price index: a hybrid model integrating LSTM with multiple GARCH-type models. Expert Syst. Appl. **103**, 25–37 (2018)

22. Kim, J.M., Jung, H.: Time series forecasting using functional partial least square regression with stochastic volatility, GARCH, and exponential smoothing. J. Forecast. **37**(3), 269–280 (2018)

23. Lipton, Z.C., Berkowitz, J., Elkan, C.: A critical review of recurrent neural networks for sequence learning. arXiv preprint arXiv:1506.00019 (2015)

24. Parkinson, M.: The extreme value method for estimating the variance of the rate of return. J. Bus. **53**(1), 61–65 (1980)

25. Peng, Y., Albuquerque, P.H.M., de Sá, J.M.C., Padula, A.J.A., Montenegro, M.R.: The best of two worlds: forecasting high frequency volatility for cryptocurrencies and traditional currencies with support vector regression. Expert Syst. Appl. **97**, 177–192 (2018)

26. Petneházi, G., Gáll, J.: Exploring the predictability of range-based volatility estimators using rnns. arXiv preprint arXiv:1803.07152 (2018)

27. Poon, S.H., Granger, C.W.: Forecasting volatility in financial markets: a review. J. Econ. Lit. **41**(2), 478–539 (2003)

28. Stock, J., Watson, M.: Dynamic factor models. In: Clements, M., Hendry, D. (eds.) Oxford Handbook of Economic Forecasting. Oxford University Press, Oxford (2010)

29. Tashman, L.J.: Out-of-sample tests of forecasting accuracy: an analysis and review. Int. J. Forecast. **16**(4), 437–450 (2000). The M3-Competition

30. Tsay, R.S.: Analysis of Financial Time Series, vol. 543. Wiley, Hoboken (2005)
31. Vincent, P., Larochelle, H., Lajoie, I., Bengio, Y., Manzagol, P.A.: Stacked denoising autoencoders: learning useful representations in a deep network with a local denoising criterion. J. Mach. Learn. Res. **11**, 3371–3408 (2010)
32. Walther, T., Klein, T.: Exogenous drivers of cryptocurrency volatility - a mixed data sampling approach to forecasting (2018). https://doi.org/10.2139/ssrn.3192474
33. Yu, S.L., Li, Z.: Forecasting stock price index volatility with LSTM deep neural network. In: Tavana, M., Patnaik, S. (eds.) Recent Developments in Data Science and Business Analytics. SPBE, pp. 265–272. Springer, Cham (2018). https://doi.org/10.1007/978-3-319-72745-5_29

Calibrating the Mean-Reversion Parameter in the Hull-White Model Using Neural Networks

Georgios Moysiadis[1]([✉]), Ioannis Anagnostou[1,2], and Drona Kandhai[1,2]

[1] Quantitative Analytics, ING Bank,
Foppingadreef 7, 1102BD Amsterdam, Netherlands
georgios.moysiadis@ing.com
[2] Computational Science Lab, University of Amsterdam,
Science Park 904, 1098XH Amsterdam, Netherlands

Abstract. Interest rate models are widely used for simulations of interest rate movements and pricing of interest rate derivatives. We focus on the Hull-White model, for which we develop a technique for calibrating the speed of mean reversion. We examine the theoretical time-dependent version of mean reversion function and propose a neural network approach to perform the calibration based solely on historical interest rate data. The experiments indicate the suitability of depth-wise convolution and provide evidence for the advantages of neural network approach over existing methodologies. The proposed models produce mean reversion comparable to rolling-window linear regression's results, allowing for greater flexibility while being less sensitive to market turbulence.

Keywords: Neural networks · Time-dependent mean-reversion · Calibration · Interest rate models · Hull-White model

1 Introduction

Stochastic models for the evolution of interest rates are a key component of financial risk management and the pricing of interest rate derivatives. The value of these products is enormous, with the outstanding notional amount currently being in the order of hundreds of trillions of US dollars [1]. One of the most widely used interest rate models is the Hull-White model, introduced in [2].

As with every computational model, the performance of the Hull-White model is significantly affected by its parameters and an improper calibration may lead to predictive inconsistencies. Specifically, the speed of mean-reversion can influence notably its results; a small value may produce more trending simulation paths, while a larger value can result in steady evolution of the interest

---

Disclaimer: The opinions expressed in this work are solely those of the authors and do not represent in anyway those of their current and past employers.

C. Alzate et al. (Eds.): MIDAS 2018/PAP 2018, LNAI 11054, pp. 23–36, 2019.
https://doi.org/10.1007/978-3-030-13463-1_2

rate. If the calibrated value does not reflect the actual market conditions, this may result in unrealistic estimates of risk exposure.

Neural networks have been proposed for the calibration of the speed of mean-reversion and volatility [3,4], as they are able to learn more complicated structures and associations in comparison to linear models. The existence of such structures is apparent, considering the relations described by the term structure of interest rate. Using neural networks to estimate variables for explicit computational models enables the experimentation with more complex and larger datasets, which can ultimately improve their performance.

The inability of simple models to exploit complex associations is perceived as an opportunity to expand the current body of research. We contribute a novel calibration method, with neural models that handle multiple historical input points of several parallel time series. Given the wide use of interest rate models and the data limitations, a future extension of neural models to exploit common features from several datasets is deemed crucial.

The rest of this paper is organized as follows. In Sect. 2 we summarize the basic features of the underlying models and methodologies that will be used. Additionally we outline several approaches related to mean-reversion calibration. In Sect. 3 we study the Hull-White model. In Sect. 4 we discuss the proposed neural network approach and the evaluation. Finally, in Sect. 5 we present our results and in Sect. 6 we summarize our findings.

2 Background and Related Work

2.1 Hull-White Interest Rate Model

The model we consider, describes interest rate movements driven by only a single source of risk, one source of uncertainty, hence it is one-factor model. This translates in mathematical terms having only one factor driven by a stochastic process. Apart from the stochastic term, the models are defined under the assumption that the future interest rate is a function of the current rates and that their movement is mean reverting. The first model to introduce the mean reverting behaviour of interest rate was proposed by Vasicek [5]. The Hull-White model is considered its extension and its SDE reads:

$$dr(t) = (\theta(t) - \alpha r(t))dt + \sigma dW(t) \tag{1}$$

where θ stands for the long-term mean, α the mean reversion, σ the volatility parameter and W the stochastic factor, a Wiener process. Calibrating the model refers to the process of determining the parameters α and σ based on historical data. $\theta(t)$ is generally selected so that the model fits the initial term structure using the instantaneous forward rate. However, its calculation involves both the volatility and mean-reversion, increasing the complexity when both are time-dependent functions.

Studying the Eq. (1) we observe the direct dependence on previous instances of interest rate which become even more obvious if we consider that θ indirectly relies on the term structure of interest rate as well. Clearly this model incorporates both temporal patterns, expressed as temporal dependencies and the market's current expectations, while the mean reversion term suggests a cyclical behaviour, also observed in many other financial indicators.

The concept of mean reversion suggests that the interest rates cannot increase indefinitely, like stocks, but they tend to move around a mean [6], as they are limited by economic and political factors. There is more than one definition of mean reversion, varying by model and scope. Mean reversion can be defined by historical floors and peaks or by the autocorrelation of the average return of an asset [7]. In the Vasicek model family, it is defined against the long term mean value towards which the rate is moving with a certain speed.

2.2 Neural Networks

Neural networks are devised as a computational equivalent of the human brain; Every neuron is a computation node that applies a non-linear function f (e.g. sigmoid), which applies a mathematical operation on the input and the connecting weights. These nodes, like in the human brain, are interconnected. They form layers that move information forward from one layer of neurons to the next. The neurons of each layer are not connected, allowing communication only with previous and succeeding layers.

Neural networks have evolved both technically and theoretically. The latest theories that are studied propose that real life data forms lower-dimensional manifolds in its embedding space [8]. For example, a set of images of a hand-written letters that include size, rotation and other transformations can be mapped to a lower dimensional space. In this manifold, the distinct variations of the same letter are topologically close. In this context, neural networks are able to learn such manifolds from the given high dimensional datasets.

This property is largely exploited in computer vision, where transformation invariance is essential for most applications. Specific modules that are formed in this weight-learning schema, such as convolutional layers, have scaled up the performance of neural networks in problems such as tracking or image recognition. It is not unusual that handcrafted models are constantly updated or replaced by neural networks.

These qualities are also applied in financial time-series for modelling and prediction tasks [9]. Models with convolutional architectures extract temporal patterns as shapes, while models with recurrent architectures learn correlations through time points. Recurrent neural networks, extend the feed-forward paradigm allowing information to move both ways. The output of a layer is connected back to its input with the appropriate trainable weights, while feeding the results to the next layer as well.

2.3 Related Work

Currently, practitioners are calibrating the Hull-White model with a variety of methods: generic global optimizers [10], Jamshidian decomposition and swap market model approximation [11], but arguably the most popular is linear regression [12]. The simplicity of the model, together with the flexibility that can be achieved by tweaking the length of the input window, make linear regression able to approximate both the market reality and one's expectations. The terms of the Hull-White model are reproduced by a linear model, given $t > s$:

$$r(t) = \hat{\alpha} r(s) + \hat{\theta} + \epsilon(t) \tag{2}$$

where $r(t)$ and $r(s)$ are the interest for the respective time-points t, s, $\hat{\theta}, \hat{\alpha}$ the historically implied long term-mean and speed of mean-reversion, and ϵ the measured error.

The trained variables are translated back to the Hull-White parameters:

$$\hat{\alpha} = e^{-\alpha(t-s)} \Rightarrow \alpha = \frac{-\ln\hat{\alpha}}{t-s}$$

$$\hat{\theta} = \frac{\theta}{\alpha}\left(1 - e^{-\alpha(t-s)}\right) \Rightarrow \theta = \frac{\alpha\hat{\theta}}{\left(1 - e^{-\alpha(t-s)}\right)} \tag{3}$$

$$sd_\epsilon = \sigma\sqrt{\frac{1 - e^{-2\alpha(t-s)}}{2\alpha}} \Rightarrow \sigma = sd_\epsilon\sqrt{\frac{-2\ln\hat{\alpha}}{(1-\hat{\alpha}^2)(t-s)}}$$

Although machine learning and neural networks have been widely used in finance for stock prediction [13], volatility modelling [14], currency exchange rate movement [15] and many more, only a few attempts have been made to address the calibration problem, and specifically, the estimation of theoretically consistent speed of mean-reversion.

Neural networks are utilized to calibrate the time-dependent mean-reversion of the Ornstein-Uhlenbeck process used for temperature derivatives in [16]. The neural network, as the approximator γ, is trained to predict the temperature of the next day. In this way, it is incorporating the dynamics of the process without explicit parameterization. Then by calculating the derivative with respect to the input of the network, they yield the value of mean-reversion. Starting from a simplified discretized version of Ornstein-Uhlenbeck process for $dt = 1$, using the paper's notation:

$$\hat{T}(t+1) = \alpha\hat{T}(t) + e(t)$$
$$\hat{T}(t+1) = \gamma(\hat{T}(t)) + e(t)$$

where $\hat{T}$ denotes the pre-processed temperature data at time t, α the simplified form of speed of mean reversion, σ the volatility and $e(t)$ the differential of the stochastic term[1]. Then computing the derivative, we calculate the time-dependent mean reversion value:

$$\alpha(t) = \frac{d\hat{T}(t+1)}{d\hat{T}(t)} = d\gamma/d\hat{T} \tag{4}$$

---

[1] Find the full expressions and derivation in [16].

3   Mean Reversion in the Hull-White Model

We have seen that mean reversion calculation can be approached in different ways depending on the underlying model. In the Vasicek model, the mean reversion parameter is assumed to be constant for certain period of time. This assumption is lifted in more generic versions, accepting it as a function of time. Under Hull-White, when calibrating with linear regression, α is re-calculated periodically on historical data as the day-to-day changes are not significant for sufficiently long historical data. This approximation results in values usually within the interval (0.01–0.1) [11]. It is common among practitioners to set alpha by hand, based on their experience and current view of the market.

Consider the generic Hull-White formula:

$$dr(t) = (\theta(t) - \alpha r(t))dt + \sigma dW(t) \tag{5}$$

Applying Ito's lemma for some $s < t$ yields:

$$r(t) = r(s)e^{-\alpha(t-s)} + \frac{\theta}{\alpha}\left(1 - e^{-\alpha(t-s)}\right) + \sigma e^{-\alpha(t-s)}\int_s^t e^{\alpha(t-u)}dW(u) \tag{6}$$

The discretized form of Eq. (5) for $dt = \delta t$ is:

$$r(t + \delta t) \approx \theta(t)\delta t + r(t)(1 - \alpha\delta t) + \sigma\epsilon(t) \tag{7}$$

where $\epsilon(t)$ denotes a value sampled from a Gaussian distribution with mean 0 and variance δt. To avoid notation abuse, this discretized approximation will be written as an equation in the following sections. Hull defined the expression for θ with constant α and σ as:

$$\theta(t) = F_t(0, t) + \alpha f(0, t) + \frac{\sigma^2}{\alpha}\left(1 - e^{-\alpha t)}\right) \tag{8}$$

where $F_t(0, t)$ is the derivative with respect to time t of $f(0, t)$, which denotes the instantaneous forward rate at maturity t as seen at time zero. Similar to [6], in our dataset the last term of the expression is fairly small and can be ignored.

4   Methodology

Our main approach for mean reversion calculation, is based on the assumption that historical rates can explain the future movement incorporating the sense of long-term or period mean value. The evolution of the yield curve is exploited in order to learn latent temporal patterns. This is achieved by training a generic function approximator that learns to predict the next-day interest rate. Starting from Hull-White model (Eq. (1)), by discretizing similar to (7) yields:

$$dr(t) = (\theta(t) - \alpha r(t))dt + \sigma dW(t)$$
$$r(t + \delta t) = \theta(t)\delta t + r(t)(1 - \alpha\delta t) + \sigma\epsilon(t) \tag{9}$$

Let $\delta t = 1$ and $\kappa = (1 - \alpha)$

$$r(t+1) = \theta(t) + \kappa r(t) + \sigma \epsilon(t) \tag{10}$$

We train a neural network with input r as the function approximator to learn the generalized version of (10) expressed as:

$$r(t+1) = \gamma(r(t)) + e(t) \tag{11}$$

where $e(t)$ denotes the measured error. Then by calculating the derivative with respect to the input of the neural network, we can compute the values of the time function $\kappa(t)$ as:

$$\kappa(t) = \frac{dr(t+1)}{dr(t)} = \frac{d\gamma}{dr} \tag{12}$$

Neural networks are constructed to be fully differentiable functions and their derivation is a well defined procedure, which can be found in [17].

4.1   Can Theta Be Replaced?

Notice that in Eq. (11), θ is not an input of γ, under the assumption that since the long term mean should be based on the current and historical interest rates, only them are needed as input to the network, which will infer the implied relations. In terms of [16], by not including θ, the equivalent preprocessing of the data is omitted. However, under the Hull-White model long term mean relies on the forward rate which describes more complex relation between $r(t+1)$ and $r(t)$ that is not modeled explicitly in the network. Removing this information from the neural model makes the approximation to have a looser connection to the assumptions of the Hull-White model; in principle, it breaks the mean reverting character. To address that, we will describe two different approaches along with Eqs. (10) and (11). The first incorporates θ, calculated on top of linear regression-calibrated alpha, resulting to:

$$\hat{r}(t+1) = r(t+1) - \theta^{LR}(t) \tag{13}$$

relying on a different model, such linear regression, introduces limitations and specific assumptions. This approach is not explored in the scope of this work.

As previously mentioned, the prevailing feature in the calculation of a long term mean is the forward rate. By keeping this information, the theoretical dependence of the movement of interest rate to the market expectations is partially embodied in the training data. This is realized by subtracting the first term of θ, F_t , to offer an approximation to long term mean value:

$$\hat{r}(t+1) = r(t+1) - F_t \tag{14}$$

Using Eqs. (13) and (14) the neural network is used to learn the evolution of $\hat{r}$:

$$\hat{r}(t+1) = \gamma(r(t)) + e(t) \tag{15}$$

4.2   Mapping

In Eq. (3) mappings from linear regression parameters to Hull-White were provided. Similarly, the calculated values from neural networks need to be transformed to be usable within the interest rate model. From Eq. (12) let $\frac{dr(t+1)}{dr(t)} = \frac{d\gamma(r)}{dr} = Y(t)$. Beginning with the simple discrete form of the Hull-White model (Eq. (7)) as expressed in Eq. (12) we get:

$$Y(t) = \kappa(t)$$
$$Y(t) = 1 - \alpha(t) \tag{16}$$
$$\alpha(t) = 1 - Y(t)$$

The mapping in Eq. (16), is appropriate for networks accepting input only the rate of the previous time-step $\delta t = 1$. To be consistent with the assumption that multiple historical input points can battle high mean-reversion volatility, the mapping used should be suitable for functions regardless the distance of the sampled points. We move back to constant α and Y for simplicity, and starting with the general continuous expression of Hull-White model, we insert the simple form of Hull-White model in our derivation:

$$Y = \frac{d\left(r(t)e^{-\alpha\delta t} + \frac{\theta}{\alpha}\left(1 - e^{-\alpha\delta t}\right) + \sigma e^{-\alpha(t-s)} \int_s^t e^{\alpha(t-u)} dW(u)\right)}{dr(t)}$$

$$Y = e^{-\alpha\delta t} \tag{17}$$

$$\alpha = \frac{-ln(Y)}{\delta t}$$

The expression for α is the same as the one used for linear regression. The output of the networks define a partial solution for each historical time-step that is provided, i.e. the partial derivative with respect to each input, that is mapped to the time-dependent alpha function. In that sense, the output of the neural network is treated as the result of the procedure $Y(t)$:

$$\alpha(t) = \frac{-ln(Y(t))}{\delta t} \tag{18}$$

4.3   Evaluation and Datasets

Three interest rate yield curve datasets were used for the experiments EUR (Fig. 2), USD (Fig. 3) and GBP (Fig. 1); The EUR dataset is comprised by 3223 data points from July 19, 2005 until November 22, 2017 with 22 maturities ranging from 1–12, 18 months, plus 2–11 years, while USD dataset starts from January 14 2002 until June 6 2011, 2446 data points with 16 maturities 1–10 and 12, 15, 20, 25, 30 and 40 years. Our datasets are fairly limited as the recorded values are daily, and in order to include all the existing rate regimes, we restricted the available maturities to 22 and 16. Our GBP data consists of 891 time-points

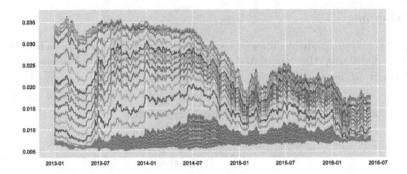

Fig. 1. Bootstrapped GBP Libor rates per maturity

from January 2, 2013 until June 1, 2016 with 44 maturities ranging from 0, 1, 7, 14 days, 1 to 24 months, 2 to 10 years and 12, 15, 20, 25, 30, 40, 50 years. The curve has been bootstrapped on top of OIS and only FRA and swap rates have been used.

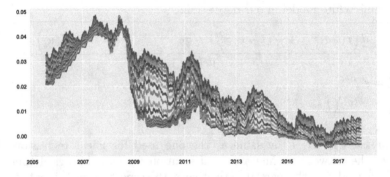

Fig. 2. European swap rate per maturity

In order to approximate the time-dependent speed of mean-reversion similarly to the linear regression method, we calculate the Pearson correlation $\in [-1, 1]$ in a rolling window manner. This way we avoid negative alpha values which cannot be used in Hull-White model. Being limited by the extent of the datasets, the window size varies from 300, 400 to 500 time points.

5   Results

Our experiments were conducted with convolution and recurrent (LSTM) based neural networks for input length of 5, 15 and 30 data points. Throughout our tests, CNNs achieved better test-set accuracy than LSTMs in all datasets, Table 1, which is not the main interest of this work, but affects the calibration procedure overall and can play a role in deciding which is the most efficient network.

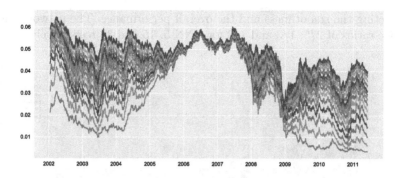

Fig. 3. USD Libor per maturity

Table 1. Average prediction error

CNN	5	15	30	LSTM	5	15	30
EUR	1.61E−04	1.08E−04	8.18E−05		3.53E−04	3.06E−04	2.96E−04
USD	1.02E−04	5.07E−05	4.92E−05		2.83E−04	2.40E−04	2.35E−04
GBP	2.10E−05	2.01E−05	2.01E−05		2.17E−05	2.24E−05	2.20E−05

The CNN approach (Fig. 5) seem to be in relative accordance with LR trends, following some of the patterns but not always agree on the levels of mean rever- sion. While there are mismatches and lags, we can spot several similarities in the movement. The evolution of alpha from CNN-5 suggests that a shorter history length results to less flexible model, at least for CNNs. Here we have three cases, with 5, 15 and 30 input length. On the lower end this behaviour is quite obvious but moving to more historical data points, the changes become stronger, resulting to higher and lower levels. In our tests, we have observed significantly different behaviours of CNNs based on the kernel_size/history_size

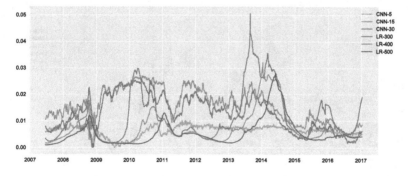

Fig. 4. EUR Mean reversion calibrated by CNN trained with Euro data

ratio affecting the smoothness and the overall performance. The networks used here, have ratios of 2/5, 1/3 and 1/3 for CNN 5, 15 and 30 respectively.

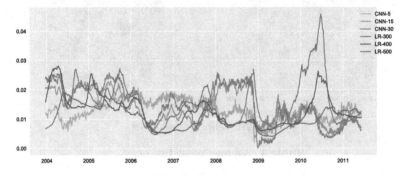

Fig. 5. USD Mean reversion calibrated by CNN trained with USD data

On the LSTM side, there is not a similar smoothness adjustment tool, which enables us to study the effect of history length easier. In Fig. 6 we see that shorter and longer history affect the outcome, but the relative movement is in close sync. In comparison to the CNN networks the levels are generally the same. We observe parallel evolution that is consistent in major changes, the two significant jumps are followed by all three networks. The first major difference from the evolution of LR (2012), seems as if the movement of LSTM is an exaggerated jump of the respective LR and CNN slopes parts. In the second (2014), we can see that the peak of the NNs coincide with the peak observed in LR-300 only, similar to CNN-30 but does not follow its level. For the USD dataset both CNN (Fig. 4) and LSTM (Fig. 7) preserve the same characteristics, while CNN-5 yields results closer to LSTMs, mostly because of the first part of the dataset (2004–2006), where lower alpha than LR and CNN (15 & 30) is reported. For the rest, LSTMs are closer to LR-500 but follow the same trends with CNNs.

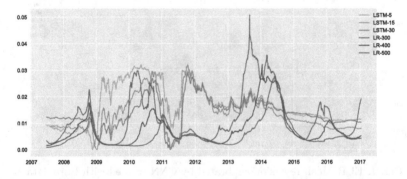

Fig. 6. EUR Mean reversion calibrated by LSTM trained with Euro data

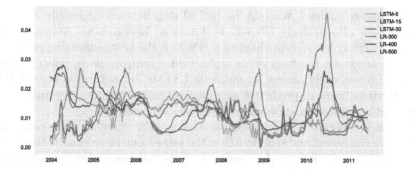

Fig. 7. USD Mean reversion calibrated by LSTM trained with USD data

In Sect. 4.1 we have discussed the role of θ in Hull-White and the significance of forward rate for its calculation. In the previous tests, the forward rate was not present in any form in the data. Following Eq. (14), we encode forward rate information by introducing the prime term of θ. The CNN-30 network is trained on this data producing the results in Fig. 8. The levels of alpha are closer to LR-500 and generally undergo smoother transitions than the simply trained CNN.

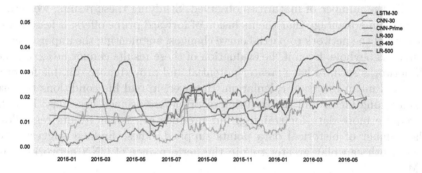

Fig. 8. CNN, LSTM, Prime calibrating GBP-Libor

Depth-wise convolution seems indeed to be suitable for parallel time-series. We observed a steady change in predictive accuracy from shorter history CNNs to longer, among the three, the network with 30-step history depth achieved the best results. However, this comparison is not simple since CNNs require certain hyper-parameter tuning, which exceeds layer count or history depth, but concerns the characteristics of the convolution layer itself; separable module, kernel size and pooling. While we focused on kernel size effects, our early tests did not favour max-pooling nor separable convolution, which led us to not further explore these options.

CNNs seem to be affected by input length and kernel size with regard to alpha calculation. 5-step networks with 2/5 kernel size ratio yield steadier results with small day-to-day changes, whereas 15 and 30 step networks generally seem to be more flexible. Particularly, CNN-15 evolution is closer to CNN-30 movement, undergoing slightly less abrupt changes. CNN-30 is the best performing network, in terms of accuracy, but also yields alpha values comparable to linear regression's more consistently among all datasets. LSTMs did not follow exactly the same pattern in terms of predictive accuracy, all three networks, regardless the history depth, resulted in approximately the same levels.

This can be observed in alpha calculations as well, where all three networks are mutually consistent and tend to follow the same slope more closely. However, the outcome in the GBP trained networks suggests that this is the case only when they are sufficiently trained.

Using the forward prime in the training phase yields smoother results, but with level mismatches with CNN-30 and LSTM-30. Observe that, overall, all networks follow the increasing trend of linear regressions, but CNN-Prime is closer and undergoes smoother changes, while big jumps are absent (Fig. 8). This is an indication that forward rate, and market expectations as an extension, can be indeed used to extract knowledge from the market and confirms Hull's indication to use the forward rate.

Generally, the behaviour of the networks is consistent in all datasets. We see that the reported mean reversion moves upward when IR experiences fast changes. The sensitivity to these changes varies depending on the network and especially the number of maturities observed. Studying these results, we recognize the main advantage that the inclusion of forward prime offers; it lessens the sensitivity of the network to partial curve changes, augmenting the importance of the curve as a whole. Even if the evaluation of these results cannot be exact, the CNN networks seem to have greater potential and proved more suitable for prediction and mean reversion calculation. However, in real life conditions the use of a smoothing factor and the inclusion of prime are deemed necessary. LSTMs may require more data for training, but produce consistent results regardless of the number of historical data points supplied. These findings suggest that history depth can play greater role in the performance of CNN networks than LSTM.

6   Conclusion

We presented a method to calibrate the speed of the mean reversion in the Hull-White model using neural networks, based only on historical interest rate data. Our results demonstrate the suitability of depth-wise convolution and provide evidence for the advantages of neural network approach over existing methodologies. The obtained mean reversion is comparable to rolling-window linear regression's results, allowing for greater flexibility while being less sensitive to market turbulence.

In the future, we would like investigate the comparison with more advanced econometric methods than linear regression, e.g. Kalman filters [18]. Another interesting direction is the understanding of learning with neural networks.

Sound model governance and model risk management frameworks are essential for financial institutions in order to comply with regulatory requirements. This creates a constrained environment for the growth of black box methods such as neural networks. Nevertheless, the latest trends in machine learning reveal that the research in the direction of opening the black box is growing [19].

Acknowledgment. This project has received funding from Sofoklis Achilopoulos foundation (http://www.safoundation.gr/) and the European Union's Horizon 2020 research and innovation programme under the Marie Sklodowska-Curie Grant Agreement no. 675044 (http://bigdatafinance.eu/), Training for Big Data in Financial Research and Risk Management.

References

1. BIS: Over-the-counter derivatives statistics. https://www.bis.org/statistics/derstats.htm. Accessed 05 Feb 2018
2. Hull, J., White, A.: Pricing interest-rate-derivative securities. Rev. Financ. Stud. **3**(4), 573–592 (1990)
3. Suarez, E.D., Aminian, F., Aminian, M.: The use of neural networks for modeling nonlinear mean reversion: measuring efficiency and integration in ADR markets. IEEE (2012)
4. Zapranis, A., Alexandridis, A.: Weather derivatives pricing: modeling the seasonal residual variance of an Ornstein-Uhlenbeck temperature process with neural networks. Neurocomputing **73**, 37–48 (2009)
5. Vasicek, O.: An equilibrium characterization of the term structure. J. Financ. Econ. **5**(2), 177–188 (1977)
6. Hull, J.: Options, Futures, and Other Derivatives. Pearson/Prentice Hall, Upper Saddle River (2006)
7. Exley, J., Mehta, S., Smith, A.: Mean reversion. In: Finance and Investment Conference, pp. 1–31. Citeseer (2004)
8. Narayanan, H., Mitter, S.: Sample complexity of testing the manifold hypothesis. In: Advances in Neural Information Processing Systems, vol. 23, Curran Associates Inc., Red Hook (2010)
9. Wei, L.-Y., Cheng, C.-H.: A hybrid recurrent neural networks model based on synthesis features to forecast the Taiwan stock market. Int. J. Innov. Comput. Inf. Control **8**(8), 5559–5571 (2012)
10. Hernandez, A.: Model calibration with neural networks. Risk.net, July 2016
11. Gurrieri, S., Nakabayashi, M., Wong, T.: Calibration methods of Hull-White model, November 2009. https://doi.org/10.2139/ssrn.1514192
12. Sepp, A.: Numerical implementation of Hull-White interest rate model: Hull-white tree vs finite differences. Technical report, Working Paper, Faculty of Mathematics and Computer Science, Institute of Mathematical Statistics, University of Tartu (2002)
13. Tsantekidis, A., Passalis, N., Tefas, A., Kanniainen, J., Gabbouj, M., Iosifidis, A.: Forecasting stock prices from the limit order book using convolutional neural networks. IEEE (2017)
14. Luo, R., Zhang, W., Xu, X., Wang, J.: A neural stochastic volatility model. arXiv preprint arXiv:1712.00504 (2017)

15. Galeshchuk, S., Mukherjee, S.: Deep networks for predicting direction of change in foreign exchange rates. Intell. Syst. Acc. Financ. Manag. **24**(4), 100–110 (2017)
16. Zapranis, A., Alexandridis, A.: Modelling the temperature time-dependent speed of mean reversion in the context of weather derivatives pricing. Appl. Math. Financ. **15**(4), 355–386 (2008)
17. LeCun, Y., Touresky, D., Hinton, G., Sejnowski, T.: A theoretical framework for back-propagation. In: Proceedings of the 1988 Connectionist Models Summer School (1988)
18. El Kolei, S., Patras, F.: Analysis, detection and correction of misspecified discrete time state space models. J. Comput. Appl. Math. **333**, 200–214 (2018)
19. Shwartz-Ziv, R., Tishby, N.: Opening the black box of deep neural networks via information. arXiv preprint arXiv:1703.00810 (2017)

Deep Factor Model

Explaining Deep Learning Decisions for Forecasting Stock Returns with Layer-Wise Relevance Propagation

Kei Nakagawa[1,3]($\boxtimes$) (iD), Takumi Uchida[3] (iD), and Tomohisa Aoshima[2,4] (iD)

[1] Nomura Asset Management Ltd., Chuo-ku, Japan
kei.nak.0315@gmail.com
[2] Fujitsu Cloud Technologies Limited, Shinjuku-ku, Japan
[3] Graduate School of Business Sciences, University of Tsukuba, Tsukuba, Japan
[4] Department of Risk Engineering, University of Tsukuba, Tsukuba, Japan

Abstract. We propose to represent a return model and risk model in a unified manner with deep learning, which is a representative model that can express a nonlinear relationship. Although deep learning performs quite well, it has significant disadvantages such as a lack of transparency and limitations to the interpretability of the prediction. This is prone to practical problems in terms of accountability. Thus, we construct a multifactor model by using interpretable deep learning. We implement deep learning as a return model to predict stock returns with various factors. Then, we present the application of layer-wise relevance propagation (LRP) to decompose attributes of the predicted return as a risk model. By applying LRP to an individual stock or a portfolio basis, we can determine which factor contributes to prediction. We call this model a deep factor model. We then perform an empirical analysis on the Japanese stock market and show that our deep factor model has better predictive capability than the traditional linear model or other machine learning methods. In addition, we illustrate which factor contributes to prediction.

Keywords: Deep factor model · Deep learning · Layer-wise relevance propagation

1 Introduction

An essential tool of quantitative portfolio management is the multifactor model. The model explains the stock returns through multiple factors. A general multifactor model in the academic finance field is sometimes used synonymously with the arbitrage pricing theory (APT) advocated by Ross [24]. The APT multifactor model includes a method of providing macroeconomic indicators a priori to explain stock returns and a method of extracting factors by factor analysis from past stock returns.

However, in practice, the Fama-French approach and the BARRA approach based on ICAPM [20] are widely used. The Fama-French or Barra multifactor

© Springer Nature Switzerland AG 2019
C. Alzate et al. (Eds.): MIDAS 2018/PAP 2018, LNAI 11054, pp. 37–50, 2019.
https://doi.org/10.1007/978-3-030-13463-1_3

models correspond to a method of finding stock returns using the attributes of individual companies such as investment valuation ratios represented by PER and PBR.

The Fama-French approach was introduced for the first time by Fama and French [9]. The Barra approach was introduced by Rosenberg [23] and was extended by Grinold and Kahn [13]. It is calculated through cross-section regression analysis since it assumes that stock returns are explained by common factors.

In addition, there are two uses of the multifactor model. It can be employed both to enhance returns and to control risk. In the first case, if one is able to predict the likely future value of a factor, a higher return can be achieved by constructing a portfolio that tilts toward "good" factors and away from "bad" ones. In this situation, the multifactor model is called a return model or an alpha model.

On the other hand, by capturing the major sources of correlation among stock returns, one can construct a well-balanced portfolio that diversifies specific risk away. This is called a risk model. There are cases where these models are confused when being discussed in the academic finance field.

For both the return model and the risk model, the relationship between the stock returns and the factors is linear in the traditional multifactor model mentioned above. By contrast, linear multifactor models have proven to be very useful tools for portfolio analysis and investment management. The assumption of a linear relationship is quite restrictive. Considering the complexity of the financial markets, it is more appropriate to assume a nonlinear relationship between the stock returns and the factors.

Therefore, in this paper, we propose to represent a return model and risk model in a unified manner with deep learning, which is a representative model that can express a nonlinear relationship. Deep learning is a state-of-the-art method for solving various challenging machine learning problems [11], e.g., image classification, natural language processing, or human action recognition. Although deep learning performs quite well, it has a significant disadvantage: a lack of transparency and limitations to the interpretability of the solution. This is prone to practical problems in terms of accountability. Because institutional investors have fiduciary duty and accountability for their customers, it is difficult for them to use black-box type machine learning technique such as deep learning. Thus, we construct a multifactor model by using interpretable deep learning.

We implement deep learning to predict stock returns with various factors as a return model. Then, we present the application of layer-wise relevance propagation (LRP [3]) to decompose attributes of the predicted return as a risk model. LRP is an inverse method that calculates the contribution of inputs to the prediction made by deep learning. LRP was originally a method for computing scores for image pixels and image regions to denote the impact of a particular image region on the prediction of a classifier for a particular test image. By applying LRP to an individual stock or a quantile portfolio, we can determine which factor contributes to prediction. We call the model a deep factor model.

We then perform an empirical analysis on the Japanese stock market and show that our deep factor model has better predictive power than the traditional linear model or other machine learning methods. In addition, we illustrate which factor contributes to prediction.

2 Related Works

Stock return predictability is one of the most important issues for investors. Hundreds of papers and factors have attempted to explain the cross section of expected returns [14, 19, 25]. Academic research has uncovered a large number of such factors, 314 according to Harvey et al. [14], with the majority being identified during the last 15 years.

The most popular factors of today (Value, Size, and Momentum) have been studied for decades as part of the academic asset pricing literature and practitioner risk factor modeling research. One of the best-known efforts in this field came from Fama and French in the early 1990s. Fama and French [9] put forward a model explaining US equity market returns with three factors: the market (based on the traditional CAPM model), the size factor (large vs. small capitalization stocks), and the value factor (low vs. high book to market). The Fama-French three-factor model, which today includes Carhart's momentum factor [6], has become a canon within the finance literature. More recently, the low risk [4] and quality factors [21] have become increasingly well accepted in the academic literature. In total, five factors are studied the most widely [15].

Conversely, the investors themselves must decide how to process and predict returns, including the selection and weighting of such factors. One way to make investment decisions is to rely upon the use of machine learning. This is a supervised learning approach that uses multiple factors to explain stock returns as input values and future stock returns as output values. Many studies on stock return predictability using machine learning have been reported. Cavalcante et al. [7] presented a review of the application of several machine learning methods in financial applications. In their survey, most of these were forecasts of stock market returns; however, forecasts of individual stock returns using the neural networks dealt with in this paper were also conducted.

In addition, Levin [18] discussed the use of multilayer feed forward neural networks for predicting a stock return with the framework of the multifactor model. To demonstrate the effectiveness of the approach, a hedged portfolio consisting of equally capitalized long and short positions was constructed, and its historical returns were benchmarked against T-bill returns and the S&P500 index. Levin achieved persistent returns with very favorable risk characteristics.

Abe and Nakayama [2] extended this model to deep learning and investigated the performance of the method in the Japanese stock market. They showed that deep neural networks generally outperform shallow neural networks, and the best networks also outperform representative machine learning models. These results indicate that deep learning has promise as a skillful machine learning method to predict stock returns in the cross section.

However, these related works are only for use as a return model, and the problem is that the viewpoint as a risk model is lacking.

3 Methodology – Deep Factor Model

3.1 Deep Learning

The fundamental machine learning problem is to find a predictor $f(x)$ of an output Y given an input X. As a form of machine learning, deep learning trains a model on data to make predictions, but it is distinguished by passing learned features of data through different layers of abstraction. Raw data is entered at the bottom level, and the desired output is produced at the top level, which is the result of learning through many levels of transformed data. Deep learning is hierarchical in the sense that in every layer, the algorithm extracts features into factors, and a deeper level's factors become the next level's features.

A deep learning architecture can be described as follows (1). We use $l \in 1,\ldots,L$ to index the layers from 1 to L, which are called hidden layers. The number of layers L represents the depth of our architecture. We let $z^{(l)}$ denote the l-th layer, and so $X = z^{(0)}$. The final output is the response Y, which can be numeric or categorical.

The explicit structure of a deep prediction rule is then

$$
\begin{aligned}
z^{(1)} &= f^{(1)}(W^{(0)}X + b^{(0)}) \\
z^{(2)} &= f^{(2)}(W^{(1)}z^{(1)} + b^{(1)}) \\
&\;\;\vdots \\
z^{(L-1)} &= f^{(L-1)}(W^{(L-2)}z^{(L-2)} + b^{(L-2)}) \\
Y &= f^{(L)}(W^{(L-1)}z^{(L-1)} + b^{(L-1)})
\end{aligned}
\tag{1}
$$

Here, $W^{(l)}$ are weight matrices, and $b^{(l)}$ are the threshold or activation levels. $z^{(l)}$ are hidden features that the algorithm extracts. Designing a good predictor depends crucially on the choice of univariate activation functions $f^{(l)}$. Commonly used activation functions are sigmoidal (e.g., $\frac{1}{(1+\exp(-x))}$, $\cosh(x)$, or $\tanh(x)$) or rectified linear units (ReLU) $\max\{x, 0\}$.

3.2 Layer-Wise Relevance Propagation

LRP is an inverse method that calculates the contribution of the prediction made by the network. The overall idea of decomposition is explained in [3]. Here, we briefly reiterate some basic concepts of LRP with a toy example (Fig. 1). Given input data x, a predicted value $f(x)$ is returned by the model denoted as function f. Suppose the network has L layers, each of which is treated as a vector with dimensionality $V(l)$, where l represents the index of layers. Then, according to the conservation principle, LRP aims to find a relevance score R_d for each vector element in layer l such that the following equation holds:

$$f(x) = \sum_{d \in V(L)} R_d^{(L)} = \cdots = \sum_{d \in V(l)} R_d^{(l)} = \cdots = \sum_{d \in V(1)} R_d^{(1)} \tag{2}$$

As we can see in the above formula (2), LRP uses the prediction score as the sum of relevance scores for the last layer of the network, and maintains this sum throughout all layers.

Figure 1 shows a simple network with six neurons. w_{ij} are weights, z_i are outputs from activation, and $R_i^{(l)}$ are relevance scores to be calculated. Then, we have the following equation:

$$\begin{aligned} f(x) &= R_6^{(3)} \\ &= R_5^{(2)} + R_4^{(2)} \\ &= R_3^{(1)} + R_2^{(1)} + R_1^{(1)} \end{aligned} \tag{3}$$

Furthermore, the conservation principle also guarantees that the inflow of relevance scores to one neuron equals the outflow of relevance scores from the same neuron. $z_{ij}^{(l,l+1)}$ is the message sent from neuron j at layer $l+1$ to neuron i at layer l. In addition, $R_d^{(l)}$ is computed using network weights according to the equation below:

$$R_i^{(l)} = \sum_j \frac{z_{ij}^{(l,l+1)}}{\sum_k z_{kj}^{(l,l+1)}} R_j^{(l+1)}, \quad z_{ij}^{(l,l+1)} = w_{ij} z_i^{(l)} \tag{4}$$

Therefore, LRP is a technique for determining which features in a particular input vector contribute most strongly to a neural network's output.

3.3   Deep Factor Model

In this paper, we propose to represent a return model and risk model in a unified manner with deep learning, which is a representative model that can express a nonlinear relationship. We call the model a deep factor model. First, we formulate a nonlinear multifactor model with deep learning as a return model.

The traditional fundamental multifactor model assumes that the stock return r_i can be described by a linear model:

$$r_i = \alpha_i + X_{i1} F_1 + \cdots + X_{iN} F_N + \varepsilon_i \tag{5}$$

where F_i are a set of factor values for stock i, X_{in} denotes the exposure to factor n, α_i is an intercept term that is assumed to be equal to a risk-free rate of return under the APT framework, and ε_i is a random term with mean zero and is assumed to be uncorrelated across other stock returns. Usually, the factor exposure X_{in} is defined by the linearity of several descriptors.

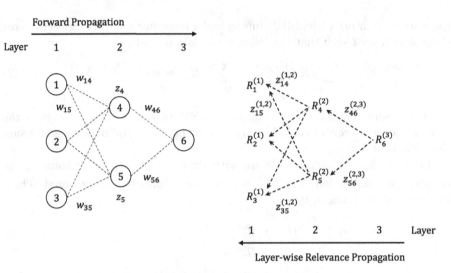

Fig. 1. LRP with toy example

While linear multifactor factor models have proven to be very effective tools for portfolio analysis and investment management, the assumption of a linear relationship is quite restrictive. Specifically, the use of linear models assumes that each factor affects the return independently. Hence, they ignore the possible interaction between different factors. Furthermore, with a linear model, the expected return of a security can grow without bound as its exposure to a factor increases.

Considering the complexity of the financial markets, it is more appropriate to assume a nonlinear relationship between the stock returns and the factors. Generalizing (5), maintaining the basic premise that the state of the world can be described by a vector of factor values and that the expected stock return is determined through its coordinates in this factor world leads to the nonlinear model:

$$r_i = \tilde{f}(X_{i1}, \ldots, X_{iN}, F_1, \ldots, F_N) + \varepsilon_i \qquad (6)$$

where $\tilde{f}$ is a nonlinear function.

The prediction task for the nonlinear model (6) is substantially more complex than that in the linear case since it requires both the estimation of future factor values as well as a determination of the unknown function $\tilde{f}$. As in a previous study [18], the task can be somewhat simplified if factor estimates are replaced with their historical means $\bar{F}_n$. Since the factor values are no longer variables, they are constants. For the nonlinear model (6), the expression can be transformed as follows:

$$\begin{aligned} r_i &= \tilde{f}(X_{i1}, \ldots, X_{iN}, \bar{F}_1, \ldots, \bar{F}_N) + \varepsilon_i \\ &= f(X_{i1}, \ldots, X_{iN}) + \varepsilon_i \end{aligned} \qquad (7)$$

where X_{in} is now the security's factor exposure at the beginning of the period over which we wish to predict. To estimate the unknown function f, a family of models needs to be selected, from which a model is to be identified. In the following, we propose modeling the relationship between factor exposures and future stock returns using a class of deep learning.

However, deep learning has significant disadvantages such as a lack of transparency and limitations to the interpretability of the solution. This is prone to practical problems in terms of accountability. Then, we present the application of LRP to decompose attributes of the predicted return as a risk model. By applying LRP to an individual stock or a quantile portfolio, we can determine which factor contributes to prediction. If you want to show the basis of the prediction for a stock return, you can calculate LRP using the inputs and outputs of the stock. In addition, in order to obtain the basis of prediction for a portfolio, calculate LRPs of the stocks included in that portfolio and take their average. Then, by aggregating the factors, you can see which factor contributed to the prediction. Figure 2 shows an overall diagram of the deep factor model.

4   Experiment on Japanese Stock Markets

4.1   Data

We prepare a dataset for TOPIX index constituents. TOPIX is a well-accepted stock market index for the Tokyo Stock Exchange (TSE) in Japan, tracking all domestic companies of the exchange's First Section. It is calculated and published by the TSE. As of March 2016, the index is composed of 1,948 constituents. The index is also often used as a benchmark for overseas institutional investors who are investing in Japanese stocks.

We use the 5 factors and 16 factor exposures listed in Table 1. These are used relatively often in practice and are studied the most widely in academia [15].

In calculating these factors, we acquire necessary data from the Nikkei Portfolio Master and Bloomberg. Factor exposures are calculated on a monthly basis (at the end of month) from December 1990 to March 2016 as input data. Stock returns with dividends are acquired on a monthly basis (at the end of month) as output data.

4.2   Model

Our problem is to find a predictor $f(x)$ of an output Y, next month's stock returns given an input X, various factors. One set of training data is shown in Table 3. In addition to the proposed deep factor model, we use a linear regression model as a baseline, and support vector regression (SVR [8]) and random forest [5] as comparison methods. The deep factor model is implemented with Tensor-Flow [1], and the comparison methods are implemented with scikit-learn [22]. Table 2 lists the details of each model.

We train all models by using the latest 60 sets of training data from the past 5 years. The models are updated by sliding one month ahead and carrying

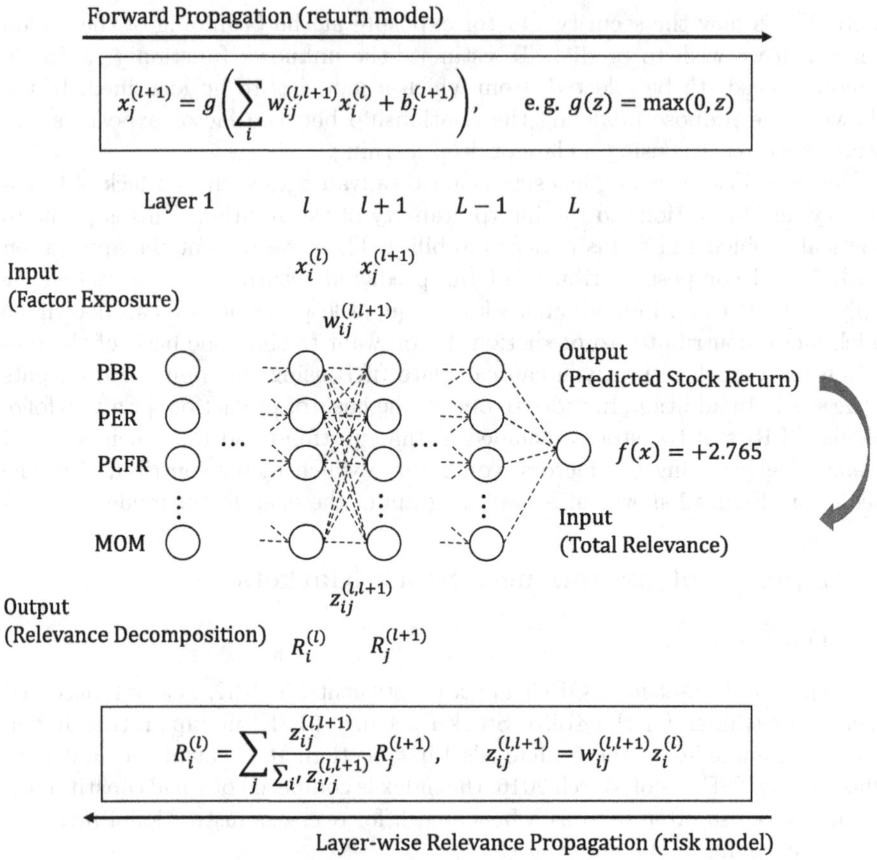

Forward Propagation (return model)

$$x_j^{(l+1)} = g\left(\sum_i w_{ij}^{(l,l+1)} x_i^{(l)} + b_j^{(l+1)}\right), \qquad \text{e.g. } g(z) = \max(0, z)$$

Fig. 2. Deep factor model

out a monthly forecast. The prediction period is 10 years, from April 2006 to March 2016 (120 months). This is because we wanted to hold a test period over 10 years including the date of Lehman shock. But, we have to check the impact of reference period choice on performance for further study. Figure 3 shows the image of our prediction framework. In order to verify the effectiveness of each method, we compare the prediction accuracy of these models and the profitability of the quintile portfolio. We construct a long/short portfolio strategy for a net-zero investment to buy top stocks and to sell bottom stocks with equal weighting in quintile portfolios. For the quintile portfolio performance, we calculate the annualized average return, risk, and Sharpe ratio. In addition, we calculate the average mean absolute error (MAE) and root mean squared error (RMSE) for the prediction period as the prediction accuracy.

Table 1. Factors and factor descriptors

Factor	Descriptor	Formula
Risk	60VOL	Standard deviation of stock returns in the past 60 months
	BETA	Regression coefficient of stock returns and market risk premium
	SKEW	Skewness of stock returns in the past 60 months
Quality	ROE	Net income/Net Assets
	ROA	Operating Profit/Total Assets
	ACCRUALS	Operating Cashflow − Operating Profit
	LEVERAGE	Total Liabilities/Total Assets
Momentum	12-1MOM	Stock returns in the past 12 months except for past month
	1MOM	Stock returns in the past month
	60MOM	Stock returns in the past 60 months
Value	PSR	Sales/Market Value
	PER	Net Income/Market Value
	PBR	Net Assets/Market Value
	PCFR	Operating Cashflow/Market Value
Size	CAP	log(Market Value)
	ILLIQ	average(Stock Returns/Trading Volume)

Table 2. Details of each method

Model		Description
Deep factor model	Model 1	The hidden layers are {80-50-10}. We use the ReLU as the activation function and Adam [16] for the optimization algorithm
	Model 2	The hidden layers are {80-80-50-50-10-10}. We use the ReLU as the activation function and Adam [16] for the optimization algorithm
Linear model		Linear models is implemented with scikit-learn with the class "sklearn.linear_model.LinearRegression" All parameters are default values in this class
SVR		Support vector regression (SVR) is implemented with scikit-learn with the class "sklearn.svm.SVR". All parameters are default values in this class
Random forest		Random Forest is implemented with scikit-learn with the class "sklearn.ensemble.RandomForestRegressor". All parameters are default values in this class

Table 3. One set of training data for March 2016.

Input: 80 dim	Output: 1 dim
Factor descriptors: 16×5 dim	Return: 1 dim
February 2016	March 2016
November 2015	
August 2015	
May 2015	
February 2015	

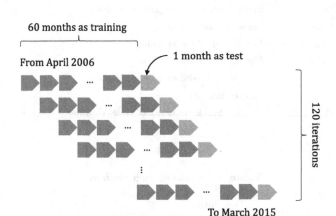

Fig. 3. Stock prediction framework.

4.3   Results

Table 4 lists the average MAE and RMSE of all years and the annualized return, volatility, and Sharpe ratio for each method. In the rows of the table, the best number appears in bold. Deep factor model 1 (shallow) has the best prediction accuracy in terms of MAE and RMSE as in the previous study [2,18]. On the other hand, deep factor model 2 (deep) is the most profitable in terms of the Sharpe Ratio. The shallow model is superior in accuracy, while the deep one is more profitable. In any case, we find that both models 1 and 2 exceed the baseline linear model, SVR, and random forest in terms of accuracy and profitability. These facts imply that the relationship between the stock returns in the financial market and the factor is nonlinear, rather than linear. In addition, a model that can capture such a nonlinear relationship is thought to be superior.

Table 4. Average MAE and RMSE of all years and annualized return, volatility, and Sharpe ratio for each method.

	Deep factor model		Linear model	SVR	Random forest
	Model 1	Model 2			
Return [%]	**10.81**	10.31	8.17	1.46	0.12
Volatility [%]	7.65	6.86	8.20	9.66	**5.43**
Sharpe ratio	1.41	**1.50**	1.00	0.15	0.02
MAE	**0.0663**	0.0669	0.0679	0.1713	0.0728
RMSE	**0.0951**	0.0953	0.0965	0.1962	0.1024

4.4 Interpretation

Here, we try to interpret the stock of the highest predicted stock return and the top quintile portfolio based on the factor using deep factor model 2 as of the last time point of February 2016. In general, the momentum factor is not very effective, but the value and size factors are effective in the Japanese stock markets. Nowadays, there is a significant trend in Japan to evaluate companies that will increase ROE over the long term because of the appearance of the Corporate Governance Code. In response to this trend, the quality factor including ROE is gaining attention. But, [17] found that both the RMW and the CMA related to our quality factor are weakly associated with the cross-sectional variations of stock returns in long term, which is significantly different from the US evidence.

Figure 4 shows which factor contributed to the prediction in percentages using LRP. The contributions of each descriptor calculated by LRP are summed for each factor and are displayed as a percentile.

We observe that the quality and value factors account for more than half of the contribution to both the stock return and quintile portfolio. The quality factor and the momentum factor are not effective in the linear multifactor model, whereas their contribution is remarkably large in the Deep Factor Model. Moreover, the contribution of the size factor is small, and it turns out that there is a widely profitable opportunity regardless of whether the stock is large or small. Figure 5 shows that these trends do not change in time series. Therefore, the Deep Factor Model is stable in terms of interpretability.

Next, we quantitatively verify the risk model by LRP. Table 5 shows the correlation coefficients between each factor and the predicted return in the top quintile portfolio. The correlation coefficients are calculated by averaging the correlation coefficients between each descriptor and the predicted return by each factor. The influence of the value and size factor differs when looking at LRP and correlation. The value factor has a large contribution to LRP and a small contribution to the correlation coefficients. The size factor has the opposite contributions. Therefore, without LRP, we could misinterpret the return factors.

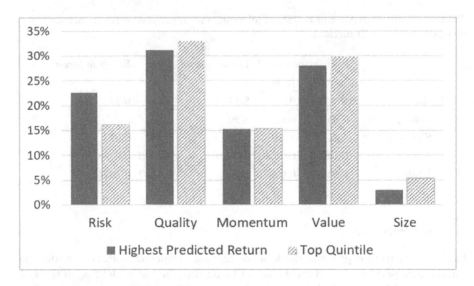

Fig. 4. Interpreting highest predicted return and top quintile portfolio based on factor using network as of last time point of February 2016

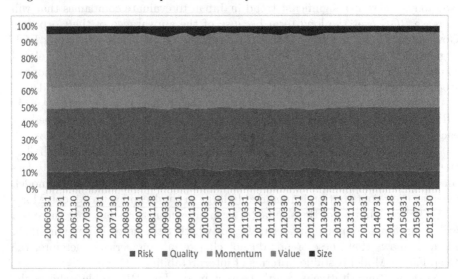

Fig. 5. Interpreting top quintile portfolio based on factor using network from April 2006 to February 2016

Table 5. Correlation coefficients between each factor and predicted return in top quintile portfolio.

	Risk	Quality	Momentum	Value	Size
Spearman	0.14	0.22	0.24	0.08	0.14
Kendall	0.10	0.15	0.17	0.06	0.10

5   Conclusion

We presented a method by which deep-learning-based models can be used for stock selection and risk decomposition. In terms of fiduciary duty and accountability for institutional investors, risk decomposition is important in practice.

Our conclusions are as follows:

- The deep factor model outperforms the linear model. This implies that the relationship between the stock returns in the financial market and the factors is nonlinear, rather than linear. The deep factor model also outperforms other machine learning methods including SVR and random forest.
- The shallow model is superior in accuracy, while the deep model is more profitable.
- Using LRP, it is possible to intuitively determine which factor contributed to prediction.

This study reports the main idea of deep factor model and initial results using Japanese stock market. We should check the stability of our model by using various stock market such as country-specific or global market [10,12].

For further study, we would like to expand our deep factor model to a model that exhibits dynamic temporal behavior for a time sequence such as RNN.

References

1. Abadi, M., et al.: TensorFlow: a system for large-scale machine learning. In: OSDI, vol. 16, pp. 265–283 (2016)
2. Abe, M., Nakayama, H.: Deep learning for forecasting stock returns in the cross-section. In: Pacific-Asia Conference on Knowledge Discovery and Data Mining (2018)
3. Bach, S., Binder, A., Montavon, G., Klauschen, F., Müller, K.R., Samek, W.: On pixel-wise explanations for non-linear classifier decisions by layer-wise relevance propagation. PloS ONE **10**(7), e0130140 (2015)
4. Blitz, D.C., Van Vliet, P.: The volatility effect: lower risk without lower return. J. Portfolio Manag (2007). Citeseer
5. Breiman, L.: Random forests. Mach. Learn. **45**(1), 5–32 (2001)
6. Carhart, M.M.: On persistence in mutual fund performance. J. Financ. **52**(1), 57–82 (1997)
7. Cavalcante, R.C., Brasileiro, R.C., Souza, V.L., Nobrega, J.P., Oliveira, A.L.: Computational intelligence and financial markets: a survey and future directions. Expert Syst. Appl. **55**, 194–211 (2016)
8. Drucker, H., Burges, C.J., Kaufman, L., Smola, A.J., Vapnik, V.: Support vector regression machines. In: Advances in Neural Information Processing Systems, pp. 155–161 (1997)
9. Fama, E.F., French, K.R.: The cross-section of expected stock returns. J. Financ. **47**(2), 427–465 (1992)
10. Fama, E.F., French, K.R.: Size, value, and momentum in international stock returns. J. Financ. Econ. **105**(3), 457–472 (2012)

11. Goodfellow, I., Bengio, Y., Courville, A., Bengio, Y.: Deep Learning, vol. 1. MIT Press, Cambridge (2016)
12. Griffin, J.M.: Are the Fama and French factors global or country specific? Rev. Financ. Stud. **15**(3), 783–803 (2002)
13. Grinold, R.C., Kahn, R.N.: Active portfolio management (2000)
14. Harvey, C.R., Liu, Y., Zhu, H.: … and the cross-section of expected returns. Rev. Financ. Stud. **29**(1), 5–68 (2016)
15. Jurczenko, E.: Risk-Based and Factor Investing. Elsevier, Amsterdam (2015)
16. Kinga, D., Adam, J.B.: A method for stochastic optimization. In: International Conference on Learning Representations (ICLR), vol. 5 (2015)
17. Kubota, K., Takehara, H.: Does the Fama and French five-factor model work well in Japan? Int. Rev. Financ. **18**(1), 137–146 (2018)
18. Levin, A.E.: Stock selection via nonlinear multi-factor models. In: Advances in Neural Information Processing Systems, pp. 966–972 (1996)
19. McLean, R.D., Pontiff, J.: Does academic research destroy stock return predictability? J. Financ. **71**(1), 5–32 (2016)
20. Merton, R.C.: An intertemporal capital asset pricing model. Econometrica: J. Econometric Soc. 867–887 (1973)
21. Novy-Marx, R.: The other side of value: the gross profitability premium. J. Financ. Econ. **108**(1), 1–28 (2013)
22. Pedregosa, F., et al.: Scikit-Learn: machine learning in Python. J. Mach. Learn. Res. **12**(Oct), 2825–2830 (2011)
23. Rosenberg, B., McKibben, W.: The prediction of systematic and specific risk in common stocks. J. Financ. Quant. Anal. **8**(2), 317–333 (1973)
24. Ross, S.A., et al.: The arbitrage theory of capital asset pricing. J. Econ. Theory **13**(3), 341–360 (1976)
25. Subrahmanyam, A.: The cross-section of expected stock returns: what have we learnt from the past twenty-five years of research? Eur. Financ. Manag. **16**(1), 27–42 (2010)

A Comparison of Neural Network Methods for Accurate Sentiment Analysis of Stock Market Tweets

Narges Tabari[✉], Armin Seyeditabari, Tanya Peddi, Mirsad Hadzikadic, and Wlodek Zadrozny

University of North Carolina at Charlotte, Charlotte, NC, USA
{nseyedit,sseyedi1,tpeddi,mirsad,wzadrozn}@uncc.edu

Abstract. Sentiment analysis of Twitter messages is a challenging task because they contain limited contextual information. Despite the popularity and significance of this task for financial institutions, models being used still lack high accuracy. Also, most of these models are not built specifically on stock market data. Therefore, there is still a need for a highly accurate model of sentiment classification that is specifically tuned and trained for stock market data.

Facing the lack of a publicly available Twitter dataset that is labeled with positive or negative sentiments, in this paper, we first introduce a dataset of 11,000 stock market tweets. This dataset was labeled manually using Amazon Mechanical Turk. Then, we report a thorough comparison of various neural network models against different baselines. We find that when using a balanced dataset of positive and negative tweets, and a unique pre-processing technique, a shallow CNN achieves the best error rate, while a shallow LSTM, with a higher number of cells, achieves the highest accuracy of 92.7% compared to baseline of 79.9% using SVM. Building on this substantial improvement in the sentiment analysis of stock market tweets, we expect to see a similar improvement in any research that investigates the relationship between social media and various aspects of finance, such as stock market prices, perceived trust in companies, and the assessment of brand value. The dataset and the software are publicly available. In our final analysis, we used the LSTM model to assign sentiment to three years of stock market tweets. Then, we applied Granger Causality in different intervals to sentiments and stock market returns to analyze the impact of social media on stock market and visa versa.

Keywords: Sentiment analysis · Neural networks · Social media · Stock market

1 Introduction

With the rise of social networks and micro-blogging, the amount of textual data on the Internet has grown rapidly, and the need to analyze it has increased along

© Springer Nature Switzerland AG 2019
C. Alzate et al. (Eds.): MIDAS 2018/PAP 2018, LNAI 11054, pp. 51–65, 2019.
https://doi.org/10.1007/978-3-030-13463-1_4

with it. Sentiment analysis has emerged as a useful and influential approach for using this data to investigate people's emotions and understand human behavior in multiple domains. For example, Bollen and Pepe [2] used social-media sentiment analysis to predict the size of markets, while Antenucci et al. [1] used it to predict unemployment rates over time.

Historically, sentiment analysis has been used to analyze longer form documents (e.g., reports, news stories, and blogs), but in the last few years, microblogging applications have seen a spike in their usage. These platforms – Twitter, Instagram, and Facebook – have rapidly become popular with professionals, celebrities, companies, and politicians, along with students, employees, and consumers of many services. The popularity of these platforms, and especially Twitter (which is text-oriented and fine-grained) provides a unique opportunity for companies and researchers to obtain a concise understanding of a single topic (e.g., the stock market) from different viewpoints.

Although social media and blogging are popular and widely used channels for discussing different topics, it is challenging to analyze their content. For example, Twitter messages generally have many misspelled words, grammatical errors, non-existent words, or unconventional writing styles. Additionally, the specific vocabulary used for analysis will depend on the topic under consideration, since the meaning and sentiment of a word can change in different contexts. For example, a word in a professional context might have positive or neutral sentiment (e.g., tax), while the same word generally has a negative sentiment in casual conversations. This prompted Loughran and Mcdonald [13] to suggest that using non-business word lists for sentiment analysis in a business context is inappropriate when using a Bag-of-Words approach.

Although many studies have concentrated on Twitter sentiment analysis in the context of the stock market, most of them either did not use a context-specific dataset, or they had low accuracy for their sentiment predictions. For example, Kolchyna et al. [10] combined lexicon-based approaches and support vector machines to classify tweets, resulting in a final accuracy of 71%. The topic of task 5 of the SemEval competition [3] was to perform fine-grained sentiment analysis on stock market tweets. Jiang et. al [7] won the first place in this competition by applying an ensemble method consisting of Random Forest, Support Vector Machine, various regression algorithms, and a combination of multiple features, such as word embeddings and lexicons. In our SemEval paper [24], we achieved an accuracy slightly lower the winning model, but with a simpler approach that used a Random Forest classifier and a revised financial lexicon from [13] as our feature set. In a recent paper, Sohangir et al. [22] evaluated regression models, data mining, and deep learning methods for sentiment analysis of financial tweets derived from StockTwits[1], and found that their CNN performed well, with an accuracy of 90.8%, while their LSTM did not perform as well, achieving an accuracy of only 69.9%.

In our work, after a precise labeling of our tweet dataset using Amazon Mechanical Turk (AMT), we applied vigorous and thorough preprocessing tech-

---

[1] www.stocktwits.com.

niques on the dataset. Then we created our baseline models, by building on our previous work [24], and then used SVM, and TF-IDF as our feature vector. Finally we thoroughly compared different Convolutional Neural Network (CNN) and Recurrent Neural Network (LSTM) with each other. We found that when using a balanced dataset of positive and negative tweets, and a specific pre-processing technique, a shallow CNN achieves the best error rate, while a shallow LSTM model, with a higher number of cells, achieves the highest accuracy of 92.7%. This is a significant improvement from our baseline or previous work in sentiment analysis in context of stock market.

Although sentiment analysis has been thoroughly studied before, we believe our work is novel in two different ways. First, there is not a publicly available annotated tweet dataset in context of stock market. Therefore, we believe that this dataset can help improve research in this area. Second, to the best of our knowledge most research on sentiment analysis in context of stock market has been studied widely using either basic machine learning classifiers or lexicon based models. Our work, on the other hand, is a one of the few thorough comparisons of neural network models that has been used in this context. And furthermore, none of previous models produced accuracy for the sentiments as high as our model.

We believe that this paper will open ways for research in a few areas measuring the impact of social media on various aspects of finance, such as stock market prices, perceived trust in companies, the assessment of brand value, and more. For instance, having a model that can predict highly accurate sentiment scores in this context, can help with the understanding of the causality analysis between social media and stock market better, or improve the prediction of stock prices using social media [2,12,13,17]. In addition, it can also be used to improve the quality of social media trust networks for stock market [18].

The outline of the paper is as follows. The dataset, the specification of how it was labeled using the Amazon Mechanical Turk, and information about labels are explained in Sect. 2. Section 3 covers the preprocessing techniques, and baseline methods. In Sect. 4, we explain all our deep learning models in details, and Sect. 5 thoroughly explains our deep learning results. In Sect. 6, first we describe our Granger Causality model and then we apply that model on the sentiments derived from Sect. 5 and stock market returns. Further, we analyze this causal analysis. And finally, we conclude our work in Sect. 7.

2   Data

Tweets were pulled from Twitter using Twitter API between 1/1/2017 and 3/31/2017. In our filters we only pulled tweets that are tweeted from a "Verified" account. A verified account on Twitter suggests that the account is a public interest and that it is authentic. An account gets verified by Twitter if the used is a distinguished person in different key interest areas, such as politics, journalism, government, music, business, and others. A tweet is considered stock related, if it contains at least one of the stock symbols of the first 100 most frequent stock symbols that were included in SemEval dataset form [24]. We were able to pull roughly 20,000 tweets in that interval using mentioned filters.

2.1 Labeling Using Amazon Mechanical Turk

The data was submitted to Amazon Mechanical Turk was asked to be labeled by 4 different workers. Snow et al. [21] suggested that 4 workers is sufficient to have enough people submitted their opinion on each tweet, and to ensure the results would be reliable. We assigned only AMT masters as our workers, meaning they have the highest performance in performing wide range of HITs (Human Intelligence Tasks). We also asked the workers to assign sentiments based on the question: "Is the tweet beneficial to the stock mentioned in tweet or not?". It was important that tweet is not labeled based on perspective of how beneficial it would be for the investor, but rather how beneficial it would be to the company itself. Each worker assigned numbers from -2 (very negative) to $+2$ (very positive) to each tweet. The inter-rater percentage agreement between sentiments assigned to each tweets by the four different workers had the lowest value of 81.9 and highest of 84.5. We considered labels 'very positive' and 'positive' as positive when calculating the inter-agreement percentage.

At the end, the average of the four sentiment was assigned to each tweet as the final sentiment. Out of 20013 tweet records submitted to AMT, we assigned neutral sentiment to a tweet if it had average score between $[-0.5, +0.5]$. We picked the sentiment positive/negative if at least half of workers labeled them positive/negative. Table 1 is a summary of the number of tweets in each category of sentiment.

One downside of this dataset was that the number of positive and negative tweets are not balanced. In order to overcome this issue, we tried many things. At the end balancing the train set by oversampling our negative tweets led to the best result. We also have tried under-sampling positive train set, but it performed worse in accuracy.

Table 1. Summary of tweets labeled by Amazon Mechanical Turk.

Range	Label assigned to tweets	Count
$[-2, -0.5]$	Negative	2082
$[-0.5, 0.5]$	Neutral	9008
$[0.5, 2]$	Positive	8386

3   Method and Models

3.1   Preprocessing

Twitter messages due to its nature of being informal text, requires a thorough preprocessing step in order to improve classifier's prediction. Twitter messages generally contain a lot of misspelled words, grammatical errors, words that does not exist, or has been written in a non-conventional way. Therefore, in our pre-processing step, we attempted to address all these issues in order to retrieve the most information possible from each tweet.

Text Substitution. We applied two different text substitutions. In our first attempt, we substitute every word that contains both number and a letter with <alphanum> tag, and all the numbers with the tag <num>. For instance, '12:30' would be replace with <num>:<num>, 'ftse100' will be replaced by <alphanum>, and '500' with <num>.

This way, all hours and measures are treated the same way. This reduces the number of non-frequent words in our vocabulary. For example, every time expression is replaced by <num>:<num>, and every price by $<num>.

Spelling Correction. In order to address the issue of misspelled words and try to retrieve as many words possible so that it can be recognizable by Word2Vec.[2] For example, we removed '-' or '.' in every word and checked whether after this operation they would be recognizable by Word2Vec. Additional preprocessing operations included:

– Removing 's'
– Changing word in 'Word1-word2' format to 'word1 word2'
– Deleting consecutive duplicate letters.
– Deleting '-' or '.' between every letter of word.

3.2   Word Embeddings

Word embeddings have been the most effective and popular feature in Natural Language Processing. The two most popular word embedding are GloVe [16] and Google's Word2Vec [14]. We used 300-dimensional pre-trained Word2Vec vectors whenever we could find a word available, and otherwise we assigned random initializations. From roughly 10,000 tokens in our vocabulary, around 600 of them was randomly initialized. It was essential for us to use pre-trained embeddings since we used to create a vocabulary in order to see if a particular word exists or not.

As future work, it would be interesting to train a new embedding model for stock market and see if that would increase the accuracy of our model.

3.3   Baseline Model

We used Amazon Mechanical Turk to manually label our stock market tweets. In order to create a baseline for our analysis, we applied on the current dataset the

---

[2] We applied Google's Word2Vec pre-trained model with 300 dimension to get word embeddings from each word.

preprocessing techniques explained before, and the same machine learning classification method and feature sets we designed for [24]. We modified Loughran's lexicon of positive and negative words [13] to be suited for stock market context, and used it to calculate the number of positive and negative words in each tweet as features. For example, 'sell' has a negative sentiment in stock market context, that has been added to Loughran's lexicon. We ultimately added around 120 new words to his list. Also, we replaced a couple of words that come together in a tweet, but has different sentiment in stock market context with one word, to be able to assign their actual sentiment. For example, 'Go down' and 'Pull back' both contain negative sentiment in stock's perceptive. Around 90 word-couples were defined specifically for this purpose. Table 2 shows the baseline for different machine learning classifiers.

Table 2. Baseline accuracy for 11,000 tweet dataset.

Classifier	Feature set	Accuracy
Random forest	[TF-IDF]	78.6%
Random forest	[TF-IDF, pos/neg count]	78.9%
Random forest	[TF-IDF, pos/neg count, Wrod-couple]	79.4%
SVM	[TF-IDF]	77.9%
SVM	[TF-IDF, pos/neg count]	**79.9%**
SVM	[TF-IDF, pos/neg count, Wrod-couple]	79.5%

4   Neural Network Models

4.1   Convolutional Neural Networks

Convolutional Neural networks (CNNs) have been shown to be useful in a variety of applications, specially in image processing. Although they have been designed originally for image processing and classification, they found their way into natural language processing. Thus models created using CNNs led to state of the art result in text classification [8,15], and specifically in classifying tweets [19,20].

Our CNN model[3] contains an input layer, in which after pre-processing, we reshape each tweet to a matrix. Then we have a convolutional layer with specific filters, and finally a max-pooling layer. Specification of each layer is described as follows:

Input Layer: CNNs originally were introduced for image classification, and by design have a fixed size input layer. Therefore, the problem with using CNNs for tweet classification is the difference in size (i.e. number of words) in tweets. To overcome this problem, we made all tweets the same size by adding padding to

---

[3] Our model, was built and modified based on a Convolutional network available at https://github.com/bernhard2202/twitter-sentiment-analysis.

shorter tweets and cutting off the longer ones to make all our tweets the same length. We set the length of tweets to 35; and among all the tweets in our data, we had only 63 tweets that had to be shortened. This way, we could represent each tweet in our dataset by a 35×300 dimensional matrix; 35 being the number of terms in each tweets, and 300 is the dimension of the representative vector in our pre-trained embeddings.

Convolutional Layer: Having our input matrix and the convolutional layer, consisting of multiple sliding window functions, the whole matrix embedding vector (word), and these convolutions slide through the matrix to generate an output with each move. For example, a filter of length 5 would go through all 35 embedding vectors (words), 5 rows at a time for 30 steps, generating 31 outputs. In our experiment we used convolutions covering three, four and five words at a time, and the output is passed to a ReLU activation function.

Max-pooling and Soft-max: Then we create a 384 dimensional vector with max-pooling on the outputs of our convolutions for each tweet (in example above each convolution creates 31 outputs for each tweet, we select the maximum and disregard all others, so we get one output for each of 384 convolutions). This output vector then will be passed to a soft-max layer to generate a normalized probability score for classification.

Training and Regularization: Stochastic optimization with cross-entropy-loss was used to train the CNN using Adam optimizer [9]. The data was divided 90% to 10% as train and development sets. After every 1000 training step the performance of the CNN on development data was evaluated and the training was stopped after eight epochs (i.e. 70k training steps) with learning rate of 1e-4. We used this learning rate, because it is low enough to make the neural network more reliable. Although, this makes the optimization process slow, it was not our concern because of our relatively small dataset. A dropout layer for convolutions was used to avoid overfitting during training. This layer disables each neuron with the probability of 0.5, resulting in a network which uses on average half the neurons in the network in each training step.

4.2 Experimenting with Recurrent Neural Networks

Recurrent neural networks have been shown to be a powerful tool in many NLP tasks such as sentiment analysis [25], machine translation [23], and speech recognition [5]. In RNNs the input is fed to the network sequentially as opposed to CNNs, where you feed the whole input into the network at once. This makes RNNs a preferred candidate for sequential data with various size inputs, such as text. They are constructed with inter-unit connections which creates a directed graph, and their internal state can be considered to be a memory which keeps track of previous states.

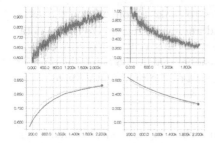

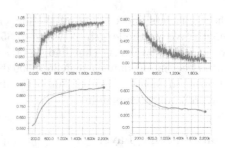

Fig. 1. Plots of accuracy and loss for each step in train and test set for best loss in CNN, from tensorboard. Top-left is the accuracy and top-right is the loss for train set. Bottom-left shows the accuracy and bottom-right shows the loss for each run in test set.

Fig. 2. Plots of accuracy and loss for each step in train and test set for best accuracy in LSTM, from tensorboard. Top-left is the accuracy and top-right is the loss for train set. Bottom-left shows the accuracy and bottom-right shows the loss for each run in test set.

An issue that arises from this design is that RNNs cannot handle long-term dependencies reliably during back propagation, resulting in vanishing or exploding gradients. This happens because the error propagates over a long distance in the network. Long Short-Term Memory (LSTM)network tries to overcome this issue by adding an explicit memory component to the network's architecture to prevent the gradients to decay very fast (and clipping large gradients prevents the exploding gradient problem). This is why we decided to try a LSTM network.

In this task, we used a network consisting an embedding layer, one layer of 128 LSTM units and a softmax layer to normalize the output. We also tried variations of this architecture: once with 256 LSTM cells, and once with two layer of 128 LSTM cells. You can see the performances for each of these architectures (along with other models) in Tables 3 and 4.

5   Results

As explained in the discussion of pre-processing, additional challenge of our dataset was the unbalanced nature of sentiments. In one experiment, we used an unbalanced test set as well as unbalanced train dataset. However, the result really jumped in accuracy when we used balanced train and test dataset. We re-sampled the negative tweets to create the same number of negative tweets as the positive ones. By doing that, our test set accuracy increased by 8% in CNN and 10% with LSTM.

Additional changes in preprocessing improved our accuracy drastically. We tried out two different preprocessing alterations. First attempt was examining the effect of removing or keeping '#' and '$' in the dataset. In all of our runs, we let these two characters remain in our dataset. The idea was that each hashtag would differentiate the word with or without these character and result in better capturing of the context. But ultimately, removing them increased the accuracy.

Table 3. Result of accuracy of different models

NN	Specification	Train	Test
CNN	Unbalanced train/test	91.5%	80.6%
CNN	Balanced train/test	89.7%	88.7%
CNN	Remove '#' and '$'	89.7%	91.6%
CNN	Unique Tag	95.9%	90.4%
LSTM	Unbalanced Train/Test	98.3%	81.6%
LSTM	Balanced Train/Test	97.9%	91.6%
LSTM	Remove '#' and '$'	91.8%	91.8%
LSTM	Unique Tag	98.4%	91.1%
LSTM	2 layer + 128 cell	83.6%	86.6%
LSTM	1 layer + 256 cell	98.4%	**92.7%**

Table 4. Result of Loss of different models

NN	Specification	Train	Test
CNN	Unbalanced train/test	0.25	0.40
CNN	Balanced train/test	0.26	0.30
CNN	Remove '#' and '$'	0.31	**0.253**
CNN	Unique tag	0.20	0.27
LSTM	Unbalanced train/test	0.07	0.68
LSTM	Balanced train/test	012	0.31
LSTM	Remove '#' and '$'	0.28	0.27
LSTM	Unique tag	0.03	0.34
LSTM	2 layer + 128 cell	0.39	0.31
LSTM	1 layer + 256 cell	0.04%	0.259

We believe this was due to the fact that our vocabulary was relatively small (10643 words), removing these characters helped with eliminating non-frequent words and reducing number of features. The effect of removing these characters can be seen in the lowest loss of 0.25 in our CNN model. Figure 1 shows the accuracy and loss for this model, for both train and test set in each step.

Second, we replaced all of our tags that have been explained in Sect. 3.1 with just one tag <num> with the same justification for removing characters. But, for both LSTM, and CNN we had slight decrease in accuracy and increase in loss.

LSTMs, in general, trained faster than CNNs, and the best accuracy was achieved when we used the higher number of LSTM cells (256) with only one layer. Our highest accuracy was 92.7% in this model, which was a significant jump from baseline. We removed both '#' and '$' from our dataset, for this model.

The 2-layer LSTM did not perform well in accuracy and loss. We believe such increase in the complexity of model would require more data for training. Figure 2 shows the accuracy, and loss for this model.

6   Comparing the Sentiments with Stock Market Returns

To begin, we downloaded the closing prices for the 100 stock ticker symbols mentioned in our labeled dataset of tweets.[4] Then, we calculated the relative daily return for each company, which is an asset's return relative to a benchmark

---

[4] Of the 100 companies mentioned, we replaced the stock symbols of companies that were owned by another with the symbol of the parent company. Specifically, we replaced $LNKD (LinkdIn) with $MSFT (Microsoft) and replaced $SCTY (Solar City) with $TSLA (Tesla). We also excluded the following companies from the list of 100 companies: VXX, GLD, SPY, GDX, SPX, WFM, EMC, APP, BRCM, and GMCR. These companies were either not currently trading, their trading data could not be found, or they were a specific index. (e.g., S&P 500).

per day. This is the preferred measure of performance for an active portfolio[5], because it is normalized, and because it a stationary time-series, a feature that is essential for most time-series analysis (and specifically, Granger causality). Stationary time-series means that they have a time-invariant mean and variance.

We used the following formula to calculate relative stock return:

$$Stock\ return = \frac{(p_1 - p_0)}{p_0}$$
$$p_0 = Initial\ stock\ price \tag{1}$$
$$p_1 = Ending\ stock\ price$$

6.1   Granger Causality Models

Granger causality (GC) is a probabilistic theory of causality [6] that determines if the information in one variable can explain another.

The advantage of this model is that it is both operational and easy to implement, but it is criticized for not actually being a model of causality (rather, it's a model of increased predictability). Critics have pointed out that even when A has been shown to Granger cause B, it does not necessarily follow that controlling A will directly influence B. Further, nor does it tell us the magnitude of the effect on B. Granger Causality is primarily used for causal notions of policy control, explanation and understanding of time-series, and in some cases, for prediction.

Formal Definition of Granger Causality: A time-series Y can be written as an autoregressive process[6], which means that the past values of Y can, in part, explain the current value of Y. Formally, an autoregressive model is defined as follows:

$$Y_t = \alpha + \sum_{i=1}^{k} \beta_j Y_{t-i} + \epsilon_t. \tag{2}$$

To define his version of causality, Granger introduced another variable X to the autoregressive model, which also has past values like Y.

$$Y_t = \alpha + \sum_{i=1}^{k} \beta_j Y_{t-i} + \sum_{j}^{k} \lambda_j X_{t-j} + \epsilon_t. \tag{3}$$

If adding X improves the prediction of current values of Y, when compared to the predictions from the autoregressive model alone, then X is said to "Granger cause" Y. Technically, Granger causality is an F-test, where the null hypothesis

---

[5] https://www.investopedia.com.

[6] An autoregressive (AR) model is a representation of a type of random process; as such, it is used to describe certain time-varying processes in nature, economics, etc. The autoregressive model specifies that the output variable depends linearly on its own previous values and on a stochastic term (an imperfectly predictable term); thus the model is in the form of a stochastic difference equation. https://en.wikipedia.org/wiki/Autoregressive_model.

is that all of the λ are equal to zero for all j. Note that you can also test the reverse case; that is, test whether Y "Granger causes" X. Both causal directions, or none, are possible. Tests for Granger causality should only be performed on stationary variables, which means that they have a time-invariant mean and variance. Specifically, this means that the variables must be $I(0)$[7] and that they can be adequately represented by a linear AR(p) process[8].

6.2   Our Granger Causality Model

Model (1):
$$RV \sim Lags(RV, LAG) + Lags(SSC, LAG) \tag{4}$$

Model (2):
$$SSC \sim Lags(SSC, LAG) + Lags(RV, LAG) \tag{5}$$

Model one determines if sentiment scores have a causal effect on stock return values, while model two determines if sentiment scores affect stock return values. In both models, the lag (LAG) is the number of days the cause precedes the effect, the return value (RV) is the calculated daily return for 83 different stocks, and the sentiment scores (SSC) are from Table 3.

6.3   Three Year Comparison of Social Media Sentiment Analysis and Stock Market Returns

In this section, we performed an in-depth causal analysis for the three stocks most commonly referred to in social media – Apple, Facebook, and Amazon – over a period of three years from 2015–2017. We used our LSTM model Table 3 to assign sentiments to an expanded Twitter dataset, which had 386,251 tweets and covered the same three year period as the stock return values. We then applied the two GC models described in 5 to find causal relationships between the sentiments and return values at five different intervals: fifteen and thirty minutes, one and three hours, and one day. For a particular interval, all of the sentiments in that interval were summed to get an aggregate score. We found causal relationship between tweet sentiments and return values for Amazon and Facebook (in both directions) at fifteen minutes, three hours, and one day. No causal relationships were found for Apple.

Looking more closely at the results of the causality analysis, we see in Tables 5 and 6 that before three hours, the value of the lag fluctuates, but at three hours

---

[7] In statistics, the order of integration, denoted I(d), of a time series is a summary statistic, which reports the minimum number of differences required to obtain a covariance-stationary series. https://en.wikipedia.org/wiki/Order_of_integration.

[8] The autocorrelation function of an AR(p) process is a sum of decaying exponentials. The simplest AR process is AR(0), which has no dependence between the terms. Only the error/innovation/noise term contributes to the output of the process, so in the figure, AR(0) corresponds to white noise. https://en.wikipedia.org/wiki/Autoregressive_model.

Table 5. F-test and P-value for three year data: sentiment causes the stock return

Stock ticker	Interval	Fvalue	Pvalue	LagNo
AMZN	1D	3.64	0.026	2
AMZN	3 h	4.33	0.013	2
AMZN	1 h	2.89	0.033	3
AMZN	30 min	2.06	0.043	7
APPL	30 min	2.08	0.034	8
FB	15 min	4	0.018	2
FB	3 h	14.74	4.30E−07	2
FB	30 min	2.31	0.04	5

Table 6. F-test and P-value for three year data: stock return causes sentiment

Stock ticker	Interval	Fvalue	Pvalue	LagNo
AMZN	15 min	4.314	0.013	2
AMZN	30 min	2.069	0.043	7
AMZN	1 h	4.59	0.01	2
AMZN	3 h	11.857	7.31E−06	2
APPL	1 h	2.395	0.014	8
FB	15 min	6.24	0.001	2
FB	1 h	2.633	0.032	4
FB	3 h	6.264	0.001	2

and one day, it stabilizes at a lag of two. We also calculated the causality weight as suggested by Geweke [4], who proved that the linear dependence of a causal model (i.e., the causality weight) can be captured by the F-measure. For both Amazon and Facebook, we found the greatest causality weight at three hours (Figs. 3 and 4). This result, along with the stabilization of the lag at three hours, suggests that we should select an interval of three hours for further analysis. The F-value and the P-value of the analysis is shown in Tables 6 and 5.

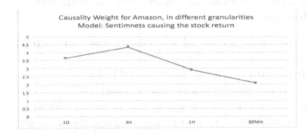

(a) Amazon shows significant causal weight on 30MIN, 1HOUR, 3HOUR and 1DAY intervals.

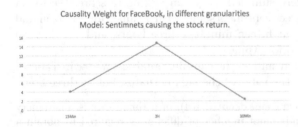

(b) Facebook shows significant causal weight on 15MIN, 3HOUR and 1DAY intervals.

Fig. 3. Statistically significant weights for model 1: sentiment causes the stock return. For both stocks, the causality weight was strongest at the 3 h time. The lowest causal weight occurred at 30 min interval.

(c) Amazon shows significant causal weight on 15MIN, 30MIN, 1HOUR, 3HOUR intervals.

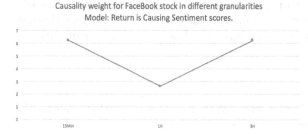

(d) Facebook shows significant causal weight on 15MIN, 1HOUR and 3HOUR intervals.

Fig. 4. Statistically significant weights for model 2: stock return causes the sentiments. For both stocks, the causality weight was strongest at the 3 h time. The lowest causal weight occurred at 30 min interval for Amazon and 1 h for Facebook.

7 Conclusion

In this paper, we first introduced a stock market related tweet dataset that has been labeled by positive or negative sentiments using Amazon Mechanical Turk. In the second part of our paper, we thoroughly compared various deep learning models, and finally introduced our LSTM model with 256 cells, which outperformed all the other models, with accuracy of 92.7%.

While this model has the best accuracy achieved in sentiment analysis of stock market tweets, there are still places for improvement. We suggest some other steps to be added to the pre-processing analysis. For example, it would be interesting to analyze the hashtag-ed words, and figure out if they are a real indicator of a subject or not (e.g. using the frequency of hashtag being mentioned in dataset). If not, they can be separated and considered to be regular words. Also, having a larger tweet dataset would help us to try out other types of deep learning models, e.g. deeper networks. Another attempt in this area could be to create domain focused word embeddings for finance.

In the final part, we analyzed the causal link between our tweet dataset, and the stock market return in different intervals. This is one of the few analyses of causality between tweets and stock prices, the other being [2,11], which has interesting result. In our analysis, we used an expanded dataset of stock return values that spanned a period of three years, from 2015 to 2017. Because we had a fine granularity of the return values and the sentiments (per minute), we partitioned both our return values and sentiment scores into five intervals: fifteen and thirty minutes, one and three hours, and one day. For each interval, we then used Granger to identify causal relationships between return values and sentiments for three companies: Apple, Facebook, and Amazon. Using Granger causality analysis at the different intervals for Amazon, Facebook, and Apple, we identified significant causal links, at a lag of three hours and one day, for Amazon and Facebook. The strongest causal weight for these two stocks occurred at a three hour lag. Importantly, the causal link existed in both directions: tweets influenced future stock market returns, and stock market returns influenced future tweets. This research can open research areas in social media impact on finance through creation of better datasets and careful analysis of other models of causality.

References

1. Antenucci, D., Cafarella, M., Levenstein, M.C., Ré, C., Shapiro, M.D.: Using social media to measure labor market flows. NBER (2014)
2. Bollen, J., Pepe, A.: modeling public mood and emotion: twitter sentiment and socio-economic phenomena, pp. 450–453 (2011)
3. Du, S., Xi, Z.: SemEval17.pdf (39), 120–125 (2016)
4. Geweke, J., Geweke, J.: Measurement of linear dependence and feedback between multiple time series measurement of linear dependence and feedback between multiple time Series **77**(378), 304–313 (2018)
5. Graves, A., Mohamed, A., Hinton, G.E.: Speech recognition with deep recurrent neural networks. CoRR abs/1303.5778 (2013). http://arxiv.org/abs/1303.5778
6. Hitchcock, C.: Probabilistic causation. In: Zalta, E.N. (ed.) The Stanford Encyclopedia of Philosophy. Metaphysics Research Lab, Stanford University, Winter 2016 (edn.) (2016)
7. Jiang, M., Lan, M., Wu, Y.: ECNU at SemEval-2017 Task 5: an ensemble of regression algorithms with effective features for fine-grained sentiment analysis in financial domain. In: Proceedings of the 11th International Workshop on Semantic Evaluation (SemEval-2017), pp. 885–890 (2017). https://doi.org/10.18653/v1/S17-2152, http://www.aclweb.org/anthology/S17-2152
8. Johnson, R., Zhang, T.: Deep pyramid convolutional neural networks for text categorization. In: Proceedings of the 55th Annual Meeting of the Association for Computational Linguistics (Volume 1: Long Papers), pp. 562–570 (2017). https://doi.org/10.18653/v1/P17-1052, http://aclweb.org/anthology/P17-1052
9. Kingma, D.P., Ba, J.: Adam: a method for stochastic optimization. CoRR abs/1412.6980 (2014). http://arxiv.org/abs/1412.6980
10. Kolchyna, O., Souza, T.T.P., Treleaven, P., Aste, T.: Twitter sentiment analysis: lexicon method, machine learning method and their combination, p. 32 (2015). http://arxiv.org/abs/1507.00955
11. Kouloumpis, E., Wilson, T., Moore, J.: Twitter sentiment analysis: the good the bad and the OMG! In: Proceedings of the Fifth International AAAI Conference on Weblogs and Social Media (ICWSM 2011), pp. 538–541 (2011). http://www.aaai.org/ocs/index.php/ICWSM/ICWSM11/paper/download/2857/3251?iframe=true&width=90%25&height=90%25

12. Lillo, F., Miccichè, S., Tumminello, M., Piilo, J.: How news affect the trading behavior of different categories of investors in a financial market. Papers.Ssrn.Com (April), 30 (2012). https://doi.org/10.1080/14697688.2014.931593, http://papers.ssrn.com/sol3/papers.cfm?abstract_id=2109337

13. Loughran, T.I.M., Mcdonald, B.: When is a liability not a liability? Textual analysis, dictionaries, and 10-Ks. J. Financ. (2010, forthcoming)

14. Mikolov, T., Chen, K., Corrado, G., Dean, J.: Efficient estimation of word representations in vector space. CoRR abs/1301.3781 (2013). http://arxiv.org/abs/1301.3781

15. Mikolov, T., Chen, K., Corrado, G., Dean, J.: Efficient estimation of word representations in vector space, pp. 1–12 (2013). https://doi.org/10.1162/153244303322533223, http://arxiv.org/abs/1301.3781

16. Pennington, J., Socher, R., Manning, C.D.: GloVe: global vectors for word representation. In: Proceedings of the 2014 Conference on Empirical Methods in Natural Language Processing, pp. 1532–1543 (2014). https://doi.org/10.3115/v1/D14-1162

17. Ranco, G., Aleksovski, D., Caldarelli, G., Grčar, M., Mozetič, I.: The effects of twitter sentiment on stock price returns. PLoS One **10**(9), 1–21 (2015). https://doi.org/10.1371/journal.pone.0138441

18. Ruan, Y., Durresi, A., Alfantoukh, L.: Using twitter trust network for stock market analysis. Knowl.-Based Syst. **145**, 207–218 (2018). https://doi.org/10.1016/j.knosys.2018.01.016, http://www.sciencedirect.com/science/article/pii/S0950705118300248

19. dos Santos, C.N., Gatti, M.: Deep Convolutional Neural Networks for Sentiment Analysis of Short Texts. Coling-2014 pp. 69–78 (2014)

20. Severyn, A., Moschitti, A.: Twitter sentiment analysis with deep convolutional neural networks. In: Proceedings of the 38th International ACM SIGIR Conference on Research and Development in Information Retrieval - SIGIR 2015, pp. 959–962 (2015). https://doi.org/10.1145/2766462.2767830, http://dl.acm.org/citation.cfm?doid=2766462.2767830

21. Snow, R., Connor, B.O., Jurafsky, D., Ng, A.Y., Labs, D., St, C.: Cheap and fast - but is it good? Evaluating non-expert annotations for natural language tasks, pp. 254–263, October 2008

22. Sohangir, S., Wang, D., Pomeranets, A., Khoshgoftaar, T.M.: Big data: deep learning for financial sentiment analysis. J. Big Data **5**(1) (2018). https://doi.org/10.1186/s40537-017-0111-6

23. Sutskever, I., Vinyals, O., Le, Q.V.: Sequence to sequence learning with neural networks. CoRR abs/1409.3215 (2014). http://arxiv.org/abs/1409.3215

24. Tabari, N., Seyeditabari, A., Zadrozny, W., Tabari, N.: SentiHeros at SemEval-2017 task 5: an application of sentiment analysis on financial tweets, pp. 857–860 (2017)

25. Zhao, X., Wang, C., Yang, Z., Zhang, Y., Yuan, X.: Online news emotion prediction with bidirectional LSTM, pp. 238–250 (2016)

A Progressive Resampling Algorithm for Finding Very Sparse Investment Portfolios

Marko Hassinen and Antti Ukkonen[✉][iD]

Department of Computer Science, University of Helsinki, Helsinki, Finland
marko.ilmari.hassinen@gmail.com, antti.ukkonen@helsinki.fi

Abstract. The mean-variance framework by Markowitz is a classical approach to portfolio selection. Earlier work has shown that the basic Markowitz portfolios obtained by solving a quadratic program tend to have poor out-of-sample performance. These issues have been addressed by devising sparse variants of Markowitz portfolios in which the number of active positions is reduced either by applying a no-short-selling constraint or L1-regularisation. In this work we consider a combinatorial approach for finding sparse portfolios, which we call naive k-portfolios, that allocate available capital uniformly on a fixed number of k assets, and only take long positions. We present a novel randomised algorithm, progressive resampling, that efficiently finds such portfolios, and compare this with a number of well-known portfolio selection strategies using public stock price data. We find that naive k-portfolios can be a viable alternative to L1-regularisation when constructing sparse portfolios.

Keywords: Portfolio selection · Randomized algorithms · Sparse models

1 Introduction

We consider a simple and novel solution to the problem of portfolio selection with non-fixed income securities. Our work is situated within the mean-variance framework popularised by Markowitz [16]. At the core of the mean-variance approach is to view the *return* of a portfolio as a random variable, and consider both its *expectation* and *variance*. In short, of all portfolios having the same expected return, a risk averse investor should choose the one with lowest variance (of returns). An increase in risk should thus always be accompanied with an increase in expected return, and vice versa.

While the Markowitz model in its basic form is a principled approach to portfolio selection, it does have some well-known downsides.

1. A solution to the Markowitz model often produces a portfolio where the investor must take a position on (almost) every available asset. Constructing (and later updating) the portfolio may thus require the investor to carry out a large number of transactions, which may be undesirable.

© Springer Nature Switzerland AG 2019
C. Alzate et al. (Eds.): MIDAS 2018/PAP 2018, LNAI 11054, pp. 66–80, 2019.
https://doi.org/10.1007/978-3-030-13463-1_5

2. Without additional constraints, the resulting portfolio may require taking a short position[1] on some of the assets, which is usually not possible for small investors, and might be considered as risky also in general.
3. Constructing the Markowitz model requires estimating a possibly very large covariance matrix from limited data (the historical asset prices). As a consequence the resulting portfolio may be prone to "overfitting" in the same way as complex machine learning models tend to overfit when training data is limited. That is, while the chosen portfolio is optimal given historical returns, it may generalise poorly to unknown future fluctuations in asset prices [2].

A possible approach to mitigate the problems above is to build a *sparse portfolio*, where the number of active positions is limited, and short selling is explicitly prohibited. It has been observed that sparse portfolios without short selling can be more stable [5,12], and they are also easier for small investors to construct due to their smaller size. Methods for constructing sparse portfolios are usually based on applying L1-regularisation within the Markowitz framework, see e.g. [5,6,9,20]. This approach is mathematically well-motivated, but it can be tricky to implement in practice, especially if it is important to impose strict constraints on the size of the resulting portfolio while simultaneously avoid short selling. For example, the approach in [5] finds a sparse portfolio that only takes long positions, but the exact number of positions taken can not be controlled by the investor.

Furthermore, it is not obvious that portfolio selection always benefits from explicit optimisation, i.e., finding the best possible portfolio given historical data. A *naive portfolio* that simply allocates a uniform amount of capital on every possible asset (and only allows long positions), can be very difficult to beat by more complex methods [7] because of the overfitting problem mentioned above.

Our Approach: In this work *we combine sparsity and naive allocation* to devise a very simple method for building sparse portfolios without short selling in which the number of active positions, expected return, and risk can be easily controlled by the investor. This basic idea as such is not new, combinatorial approaches (e.g. based on semidefinite programming) to portfolio selection have been proposed earlier [10,21]. However, we view portfolio selection as a combinatorial search problem instead of a numerical optimisation task. Also, our randomised search algorithm is novel, as well as is our relaxed notion of "optimality" of the portfolio.

Our contributions are defining the naive k-portfolio selection problem, and presenting the Progressive Resampling algorithm for solving it[2]. We also present two experiments on stock market data to illustrate the method.

---

[1] "Short selling" means that the investor borrows the asset and immediately sells it on the market. If the price of the asset decreases after some time, the investor can buy the asset back at a lower price and keep the difference before returning the asset to the lender. Short selling is essentially a bet that the price of an asset will decrease, while simply buying the asset (or, taking a "long position") is a bet that the assets price will increase.

[2] Variants of these have been introduced in the Master's thesis of the first author.

2 Background and Problem Statement

We first discuss basic definitions of the mean-variance framework (please see e.g. [15] for a more in-depth discussion), then briefly introduce relevant background related to L1-regularised solutions, and conclude by presenting our novel problem of naive k-portfolio selection.

2.1 The Markowitz Model and Its L1 Regularised Variant

Let $A = \{a_1, \ldots, a_n\}$ denote a set of n *assets*, and denote by $p_i(t)$ the *price* of asset a_i at time t. In practice the $p_i(t)$ could refer to the daily (or weekly, monthly) closing price at some stock exchange, for example. The *return* of asset a_i at time t is denoted $r_i(t)$. Returns are defined as the ratio between the price of the asset at two successive time intervals, i.e., $r_i(t) = p_i(t)/p_i(t-1)$. A *portfolio* is represented by a weight vector $\mathbf{w} \in \mathbb{R}^n$, where w_i is the weight of asset a_i in the portfolio, and $\sum_{i=1}^{n} w_i = 1$. The return of portfolio $\mathbf{w}$ at time t is simply the weighted sum of asset returns: $r_{\mathbf{w}}(t) = \sum_{i=1}^{n} w_i r_i(t)$.

Since in a non-fixed income setting future asset prices are unknown, the $r_i(t) = r_i$ are considered to be random variables. Let $\mathbf{r} \in \mathbb{R}^n$ denote the vector of expected returns for every asset, and denote by $\mathbf{\Sigma}$ the matrix of covariances between all pairs of assets in A. Both $\mathbf{r}$ and $\mathbf{\Sigma}$ are estimated from historical prices. In this paper we consider their standard estimators. The *expected return* and *variance* of portfolio $\mathbf{w}$ are now given by $\mathbf{w}^\mathsf{T}\mathbf{r}$ and $\mathbf{w}^\mathsf{T}\mathbf{\Sigma}\mathbf{w}$, respectively.

Given $\mathbf{r}$ and $\mathbf{\Sigma}$, the original Markowitz model is defined by the following quadratic program that minimizes portfolio risk for a given return level μ_0:

$$\min_{\mathbf{w}} \; \mathbf{w}^\mathsf{T}\mathbf{\Sigma}\mathbf{w} \tag{1}$$

$$\text{st.} \quad \mathbf{w}^\mathsf{T}\mathbf{r} = \mu_0 \tag{2}$$

$$\sum_{i \in A} w_i = 1, \tag{3}$$

where μ_0 is a user-specified expected return given by the investor, and $\mathbf{1}$ is the vector of all ones. It is straightforward to show (see e.g. [5]) that this is equivalent to the least squares problem

$$\min_{\mathbf{w}} \; \|\mu_0 \mathbf{1} - \mathbf{R}\mathbf{w}\|_2^2 \tag{4}$$

$$\text{st.} \quad \mathbf{w}^\mathsf{T}\mathbf{r} = \mu_0, \tag{5}$$

$$\mathbf{w}^\mathsf{T}\mathbf{1} = 1, \tag{6}$$

where $\mathbf{R}$ is the matrix of returns over time, i.e., $R_{ti} = r_i(t)$. This least squares formulation immediately leads to standard regularisation approaches, in particular an L1-penalty term can be added in the objective function to promote sparse solutions:

$$\min_{\mathbf{w}} \left(\|\mu_0 \mathbf{1} - \mathbf{R}\mathbf{w}\|_2^2 + \tau \|\mathbf{w}\|_1 \right) \tag{7}$$

$$\text{st.} \; \mathbf{w}^\mathsf{T}\mathbf{r} = \mu_0, \tag{8}$$

$$\mathbf{w}^\mathsf{T}\mathbf{1} = 1, \tag{9}$$

where τ is a regularisation parameter. Unconstrained instances of such problems can be efficiently solved e.g. with the LARS algorithm [8]. In [5] a variant of LARS is devised that takes linear constraints, such as the ones required here, into account. However, as suggested in the Appendix of [5], the constrained L1-Regularised problem can be solved approximately with standard LARS by including the constraints in the objective function as follows:

$$\min_{\mathbf{w}} \left(\|\mu_0 \mathbf{1} - \mathbf{R}\mathbf{w}\|_2^2 + \gamma \| [1; \mu_0] - [\mathbf{1}; \mathbf{r}] \, \mathbf{w}\|_2^2 + \tau \|\mathbf{w}\|_1 \right), \qquad (10)$$

where $[1; \mu_0]$ denotes a (column) vector with the corresponding entries, and $[\mathbf{1}; \mathbf{r}]$ is a $2 \times n$ matrix where the first row contains ones, and the second row is equal to $\mathbf{r}$. The parameter γ must be set to a large enough value to avoid having $\|\mu_0 \mathbf{1} - \mathbf{R}\mathbf{w}\|_2^2$ dominate the objective function. In the experiments we compare our algorithm against this approximate approach. LARS finds a solution to Eq. 10 by adding assets one-by-one starting from zero assets. In practice the constraints are satisfied (approximately) only after a certain number of assets have been added. Hence, we always use the solution with the smallest number of active positions that satisfies $\mathbf{w}^\mathsf{T}\mathbf{1} \geq 0.9999$.

Next, observe that while the non-regularised Markowitz model (Eqs. 1–3) can be forced to produce portfolios without short-selling by adding the explicit constraint $w_i \geq 0$ for all $a_i \in A$, this can unfortunately not be done for the L1-regularised variant. If we force all $w_i \geq 0$ while keeping the constraint $\mathbf{w}^\mathsf{T}\mathbf{1} = 1$, the L1-term $\|\mathbf{w}\|_1 = \sum_{i=1}^n |w_i|$ is always equal to 1, and hence regularisation does not have any effect at all! While solutions to the regularised least-squares problem given in Eqs. 7–9 can be both sparse and non-negative, it is thus not possible to enforce non-negativity, at least not simply by adding a constraint. Also the degree of sparsity may vary: sometimes the solutions are very sparse, at other times they can be less so, depending on the return matrix $\mathbf{R}$ and how μ_0 is set.

2.2   Naive k-portfolios

We move on to present our variant of the portfolio selection problem that aims to address these issues. In particular, we want to *enforce no short-selling and have control over portfolio size*, but are willing to sacrifice (a little) both in optimality of return and risk. Basically we look for a portfolio that has a long position on *exactly k assets*, with each asset having the *same weight $1/k$*, so that the expected return of the portfolio is larger than a given threshold μ^*, and the variance of the portfolio is less than a given threshold σ^*.

Definition 1. *Let $U_k \subset \mathbb{R}^n$ denote the set of vectors $\mathbf{w}$ for which $w_i \in \{0, 1/k\}$ for all $a_i \in A$, and $\sum_{i=1}^n w_i = 1$. That is, U_k is the set of all possible portfolios of assets A that have a positive weight of $1/k$ on exactly k assets, and all other assets have zero weight. We call such portfolios naive k-portfolios.*

Definition 2. *Given return vector $\mathbf{r}$ and covariance matrix Σ, define the feasible set $\mathcal{F}(\mu, \sigma) = \{\mathbf{w} \in U_k \mid \mathbf{w}^\mathsf{T}\mathbf{r} \geq \mu \text{ and } \mathbf{w}^\mathsf{T}\Sigma\mathbf{w} \leq \sigma\}$.*

Problem 1. Given integer k, minimum return μ^* and maximum risk σ^*, find at least one portfolio $\mathbf{w} \in \mathcal{F}(\mu^*, \sigma^*)$.

Clearly Problem 1 is not an optimisation problem, but a combinatorial search task. Depending on how the investor sets the parameters μ^* and σ^*, the number of feasible portfolios varies. In our approach, all portfolios in $\mathcal{F}(\mu^*, \sigma^*)$ are considered as equally good choices. Notice also that if μ^* and σ^* are set inappropriately, the set of feasible portfolios is empty, and no solution exists to Problem 1.

3 A Progressive Resampling Algorithm

3.1 Basic Idea

Since Problem 1 only asks to find *one* portfolio that satisfies the constraints on return and risk, it may be possible to use a random search procedure. If the constraints μ^* and σ^* are loose enough, perhaps there are so many feasible naive k-portfolios that we can simply draw vectors $\mathbf{w}$ from U_k at random and it will not take too long to find one from $\mathcal{F}(\mu^*, \sigma^*)$ that satisfies the constraints. The problem becomes computationally challenging (and interesting) only when feasible portfolios are *rare*, meaning that a (uniform) random sampler is unlikely to find them in a reasonable amount of time.

To remedy this situation, we devise an algorithm that resembles importance sampling. While the idea of importance sampling in general is to accurately estimate e.g. the probabilities of rare events, we are only interested in *making one of these rare events happen*. This is done by iteratively *adjusting the sampling distribution* from which the portfolios are drawn. Consider a distribution that assigns a positive probability to portfolios in $\mathcal{F}(\mu^*, \sigma^*)$, and a probability of zero to all other portfolios. Drawing a single $\mathbf{w}$ from this ideal distribution is guaranteed to yield a solution to Problem 1! Obviously this is infeasible in practice, as setting up such an ideal distribution basically requires us to solve the problem. But this intuition serves as the basis of our algorithm. We start from a uniform sampling distribution, and then step-by-step adjust the sampler towards a distribution that is more likely to produce portfolios from $\mathcal{F}(\mu^*, \sigma^*)$, and assigns a lower probability to the infeasible portfolios.

Algorithm 1. Progressive resampling

Start with relaxed constraints $\mu^1 = \mu^*/M$ and $\sigma^1 = M\sigma^*$ (M is some suitably large number), and a uniform sampling distribution over U_k. Then on each iteration,

1. draw a random collection S of naive k-portfolios from the current distribution,

2. tighten the relaxed constraints towards μ^* and σ^* so that a fixed proportion of portfolios in S remain satisfied,

3. update the sampling distribution to assign a higher probability on portfolios in S that satisfy the updated constraints.

Repeat steps 1–3 until a portfolio that satisfies the original constraints μ^* and σ^* is found in S, or the constraints can no longer be tightened.

An outline of the approach is shown in Algorithm 1. Next we discuss steps 1–3 in detail.

Step 1: Sampling Random Portfolios

Let $\Pr[\mathbf{w}; \Theta]$ denote the probability of observing portfolio $\mathbf{w}$ from a sampling distribution parametrised by Θ. We consider distributions where this probability is expressed in terms of *asset-specific weights*. The parameters Θ are simply these asset-specific weights, i.e., $\Theta \in [0, 1]^n$ is a vector of probabilities so that $\sum_{i=1}^{n} \Theta_i = 1$.

To draw a random portfolio given Θ, we employ a standard "without replacement" subset sampling algorithm that applies the weights in Θ sequentially to draw one asset at a time. After each draw, weights of the remaining assets are normalised so that they sum up to 1. The probability of portfolio $\mathbf{w}$ given Θ is given by

$$\Pr[\mathbf{w}; \Theta] = \prod_{i:w_i>0} \Theta_i \left(\sum_\pi \prod_{j=1}^{k-1} \left(1 - \sum_{i=1}^{j} \Theta_{\pi(i)} \right) \right)^{-1}. \tag{11}$$

Above π runs over all permutations of integers $1, \ldots, k$. $\Pr[\mathbf{w}; \Theta]$ is thus equal to the product $\prod_{i:w_i>0} \Theta_i$, divided by a sum over all permutations π of the assets in the portfolio, where every π contributes a term the value of which can be thought of as inversely proportional to $\sum_{i:w_i>0} \Theta_i$. Therefore, clearly $\Pr[\mathbf{w}; \Theta]$ increases when we increase Θ_i for those assets i that have a positive weight in $\mathbf{w}$. It is also straightforward to show that if $\Theta_i = 1/n$ for all $i \in A$, or if $\Theta_i = 1/k$ for $i : w_i > 0$ and $\Theta_i = 0$ elsewhere, Eq. 11 simplifies to $\binom{n}{k}^{-1}$ and 1, respectively.

Step 2: Tightening the Constraints

Let μ^t and σ^t denote the (relaxed) constraints at the start of the t:th iteration. We tighten the constraints by moving μ^t and σ^t towards μ^* and σ^* so that a (user specified) proportion $\gamma \in [0, 1]$ of portfolios in the collection S remain satisfied. To do this, we must *increase* μ^t and *decrease* σ^t. While this could be accomplished in a number of ways, for now we propose to scale both μ^t and σ^t with the same factor δ. That is, we find a $\delta > 1$, such that $|\{S \cap \mathcal{F}(\delta\mu^t, \sigma^t/\delta)\}| = \gamma|\{S \cap \mathcal{F}(\mu^t, \sigma^t)\}|$, and then let $\mu^{t+1} = \delta\mu^t$ and $\sigma^{t+1} = \sigma^t/\delta$. The δ can be found e.g. with binary search.

Step 3: Updating the Sampling Distribution

Let Θ^t denote the asset-specific weights at the start of the t:th iteration. Note that when $t = 1$ we have $\Theta_i^1 = 1/n$ for all $i \in A$. The objective of the update is to increase the probability of those portfolios in collection S that satisfy the updated constraints μ^{t+1} and σ^{t+1} from step 2. Let $S^\mathcal{F} = S \cap \mathcal{F}(\mu^{t+1}, \sigma^{t+1})$ denote the subset of S that satisfies the updated constraints, and let $q_i = |S^\mathcal{F}|^{-1} \sum_{\mathbf{w} \in S^\mathcal{F}} w_i$. If almost all, or none of the portfolios in $S^\mathcal{F}$ contain i, q_i will be close to 1 or 0, respectively. The sampling weights for all assets i are then updated by the formula $\Theta_i^{t+1} = (1-\alpha)\Theta_i^t + \alpha q_i$, where $\alpha \in [0, 1]$ can be thought of as a "learning rate".

Stopping Condition

A trivial stopping condition is obviously a check that determines if any $\mathbf{w} \in S$ also belongs to $\mathcal{F}(\mu^*, \sigma^*)$. In addition to this, the algorithm stops whenever *in Step 2 no $\delta > 1 + \epsilon$ can be found*, where ϵ is some very small constant. This means that even if the constraints are tightened only by a very small amount, all satisfying portfolios in the current collection S become unsatisfying, and we can no longer continue with the tightening process. This could mean that $\mathcal{F}(\mu^*, \sigma^*)$ in fact is empty, and stopping without returning a solution is the correct choice. But we can not rule out that this is simply caused by the collection S missing exactly those portfolios that reside "on the way" towards $\mathcal{F}(\mu^*, \sigma^*)$. A rigorous analysis of this is left as future work, for now we just consider this as a reasonable stopping heuristic.

3.2   Why Should Progressive Resampling Work?

The algorithm aims to adjust Θ so that the probability of portfolios in $\mathcal{F}(\mu^*, \sigma^*)$ increases enough so that eventually at least one of them appears in the sample S. We sketch a brief argument that sheds some light on how the algorithm tries to achieve this.

We consider an idealised variant of the algorithm in which on each iteration the collection S is equal to U_k irrespectively of Θ^t. To simplify matters further, suppose there is *only one* portfolio in $\mathcal{F}(\mu^*, \sigma^*)$. Also, suppose that $\alpha = 1$, and observe that after tightening the constraints we thus always move Θ^{t+1} to the "center" of $\mathcal{F}(\mu^{t+1}, \sigma^{t+1})$. (It helps to think that Θ^t and the portfolios reside in the same vector space.)

Next, note that the sets $\mathcal{F}(\mu^1, \sigma^1), \mathcal{F}(\mu^2, \sigma^2), \ldots$ considered by this idealised algorithm *by construction* form a nested (possibly very long) sequence of subsets of U_k that "approaches" $\mathcal{F}(\mu^*, \sigma^*)$, meaning we have

$$U_k \supseteq \mathcal{F}(\mu^1, \sigma^1) \supset \mathcal{F}(\mu^2, \sigma^2) \supset \ldots \supset \mathcal{F}(\mu^*, \sigma^*).$$

This, together with the observation above, means that the vector Θ^t must converge towards the single feasible portfolio which will hence eventually appear in S.

The claim is that the progressive resampling algorithm approximates this idealised procedure using a "small" collection S that is resampled on each iteration from the updated distribution. This idea can be seen as akin to stochastic gradient descent, where on each iteration the gradient is computed only from a small batch rather than all of the training data.

The accuracy of this approximation depends on the size of S, as well as the parameters α and γ. The parameter α serves two purposes. First, it guarantees that we will always have a small nonzero probability on every $i \in A$, even if some asset might not have appeared in S at all. Second, since in practice $|S| \ll |U_k|$, the q_i might be a poor approximation of the center of $\mathcal{F}(\mu^{t+1}, \sigma^{t+1})$. By letting $\alpha < 1$ these issues can be mitigated to some degree. Furthermore, by setting γ in step 2 to a small (but not too small) value, we can increase the chances of finding a large enough $\delta > 1$, and thus the resampling can continue. Note that like in the idealised algorithm, the sequence of feasible sets will be nested and always contains $\mathcal{F}(\mu^*, \sigma^*)$. However, if γ is set very low, the resulting q_is are computed from very few examples, meaning they might not guide the process "in the right direction". This latter problem can be mitigated by drawing a "large enough" collection S. A more formal study of the interplay of the algorithm's parameters is left for future work.

4 Experiments

4.1 Datasets and Baseline Methods

We use two stock market datasets with daily closing prices of securities. These are (1) a Dow-Jones-30 dataset (DJ30) of 30 stocks[3], and (2) a subset of 322 SP500 stocks.

We consider the following well-known portfolio selection methods in the experiments.

1. In the **Naive portfolio** we simply set $w_i = 1/n$ for all $a_i \in A$.
2. The **Basic Markowitz** portfolio is obtained by solving $\min_w \mathbf{w}^\mathsf{T} \Sigma \mathbf{w}$ subject to $\mathbf{w}^\mathsf{T}\mathbf{r} = \mu_0$ and $\mathbf{w}^\mathsf{T}\mathbf{1} = 1$.
3. The **No-short Markowitz:** is obtained in the same way, but with the additional constraint $w_i \geq 0$ for all $a_i \in A$.
4. To obtain the **L1-Markowitz** portfolio we solve Eq. 10 with LARS, and select the solution with the smallest number of active positions that satisfies $\mathbf{w}^\mathsf{T}\mathbf{1} \geq 0.9999$. These portfolios also almost always have no short positions, but this is not explicitly guaranteed.

4.2 Experiment 1: Are There Any Reasonable Naive k-Portfolios?

The purpose of the first experiment is to study if naive k-portfolios can compete with optimised portfolios in terms of their risk/return characteristics. We first estimate daily returns and their covariances using all of the available data for both DJ30 and SP500, and find the optimal L1-Markowitz portfolio for a number of return levels with the approximate LARS algorithm. These are shown as black crosses in Fig. 1 for both DJ30 (top) and SP500 (bottom) with risk and return on the x and y-axis, respectively.

Since DJ30 has only 30 assets in total, we then enumerate *all possible* naive k-portfolios for different values of k, and then plot the *efficient frontier of these* (i.e. the Pareto-optimal naive k-portfolios). This is shown for $k \in \{5, 8, 10\}$ by the red lines in the top panel of Fig. 1. *All* naive k-portfolios for the given k reside below (and to the right) of the respective line. Clearly when there are only few assets in total, the naive k-portfolios have a substantially higher risk for a given return level than the optimal L1-Markowitz portfolios, and even if k is increased from 5 to 10, their risk is reduced only by a small amount.

---

[3] http://lib.stat.cmu.edu/datasets/DJ30-1985-2003.zip.

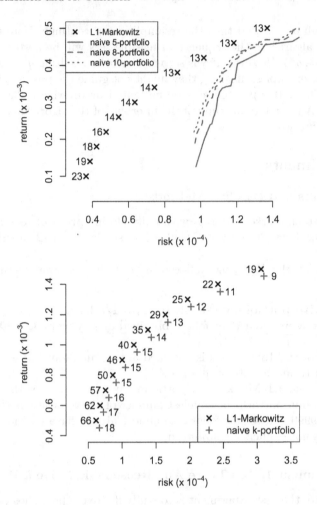

Fig. 1. Experiment 1. Naive k-portfolios compared with optimal L1-Markowitz portfolios. Numbers adjacent to the symbols indicate the number of active positions in the portfolio. *Top, DJ30:* Exact efficient frontiers of naive k-portfolios for different values of k. The L1-Markowitz portfolios have substantially lower risk than naive k-portfolios. *Bottom, SP500:* In this case there exist naive k-portfolios with comparable risk/return characteristics as the L1-Markowitz portfolios. See text for details.

For SP500 there are simply too many assets to exhaustively enumerate all possible naive k-portfolios. However, now we can use the progressive resampling algorithm to investigate if there are any naive k-portfolios that are reasonably close by to the optimal L1-Markowitz portfolios. Given return μ_{L1} and risk σ_{L1} of an L1-portfolio $\mathbf{w}_{L1}$, we set the constraints μ^* and σ^* so that $\mathbf{w}_{L1} \in \mathcal{F}(\mu^*, \sigma^*)$, by letting $\mu^* = \mu_{L1} - 5 \times 10^{-5}$ and $\sigma^* = \sigma_{L1} + 5 \times 10^{-6}$. Then, for every L1-Markowitz portfolio in the bottom panel of Fig. 1, we find the smallest naive k-portfolio that satisfies the respective μ^* and σ^*. (The smallest portfolio is found simply by running the progressive resampling algorithm with increasing values of k until a solution is found.) The red plus signs in Fig. 1 (bottom panel) show risk and return of these. We can observe that there indeed exist naive k-portfolios that are comparable to the optimal L1-Markowitz portfolios in terms of return and risk, but have a substantially smaller number of active positions, especially in the low-risk low-return part.

In summary, we find that naive k-portfolios can have comparable performance to L1-Markowitz portfolios, but this seems to hold only when the total number of assets is large enough. If the portfolio must be assembled from a set of only a few assets, L1-regularisation seems like a more sold approach. However, these observation are based on "training data". Next, we proceed to study out-of-sample performance of the different portfolio selection strategies.

4.3   Experiment 2: Out-of-Sample Performance of Naive k-Portfolios

In this experiment we only use the SP500 data, because above we observed that for DJ30 the naive k-portfolios have substantially higher excepted risk in "training data" than optimised portfolios. We use historical prices from a given *two-year period* to estimate daily asset returns and their covariances. The resulting portfolios are evaluated in terms of the Sharpe-ratio (expected return divided by its standard deviation) with prices from the *one-year period that follows* the day on which the portfolio was constructed (always June 30th, we consider the years 1998–2013). We thus repeat the portfolio selection process for a number of (overlapping) training periods to assess how the different portfolio selection strategies perform in different market conditions. (This is not to suggest that an investor should construct a new portfolio each year. Our aim is to simply see how the methods perform over time for different investors. A similar evaluation methodology was also taken e.g. in [5].)

We set the desired daily return level μ_0 so that it corresponds to a 1-year return of seven percent (7%). The progressive resampling algorithm was called with μ^* and σ^* set to replicate the performance of the L1-Markowitz portfolio by slightly relaxing its risk and return level. Also, we let $k = 15$, meaning the resulting portfolio should consist of exactly 15 assets. Given that there are 322 assets in total, this results in a very sparse portfolios. Since our algorithm is randomised and is thus likely to return a different portfolio even with the same input parameters, we run the algorithm 25 times and report the average performance of the portfolios thus obtained.

Results are shown in Table 1. The selection strategy having the highest Sharpe ratio is highlighted in **bold**. The leftmost column shows the date on which the portfolio was constructed. That is, it was trained on data from the previous two years, and evaluated on the daily returns of the 12 months that followed. We find that the naive portfolio that simply allocates all capital uniformly over all possible n assets tends to have surprisingly good out-of-sample performance. This result has been observed also earlier, see e.g. [7]. Also, the basic Markowitz model without short-selling constraints nor regularisation tends to perform poorly: in good times it can not really outperform the naive strategy,

Table 1. Out-of-sample performance on SP500 daily returns (Sharpe ratio)

	Naive	Basic Markowitz	No-short Markowitz	L1- Markowitz	Naive-15
30-06-1998	**0.08**	0.00	0.01	0.03	0.02
30-06-1999	**0.03**	−0.06	−0.03	−0.04	−0.02
30-06-2000	0.09	0.09	**0.14**	0.12	0.12
30-06-2001	0.01	−0.02	**0.05**	**0.05**	−0.05
30-06-2002	**0.02**	**0.02**	**0.02**	0.01	−0.01
30-06-2003	0.14	0.14	**0.15**	0.13	0.13
30-06-2004	0.09	0.03	**0.10**	0.09	-
30-06-2005	**0.07**	0.02	0.03	0.04	0.04
30-06-2006	0.11	0.12	0.12	**0.14**	0.11
30-06-2007	**−0.02**	−0.07	−0.03	−0.03	−0.03
30-06-2008	−0.01	−0.05	**−0.01**	−0.03	**−0.01**
30-06-2009	0.08	0.10	**0.13**	**0.13**	-
30-06-2010	0.13	0.15	0.16	**0.17**	-
30-06-2011	0.01	**0.12**	0.06	0.05	0.07
30-06-2012	**0.11**	0.04	0.08	0.06	0.07
30-06-2013	**0.13**	−0.01	0.04	0.04	0.04

and in bad times it tends to have substantially lower Sharpe ratio. No-Short Markowitz on the other hand is another good candidate.

When comparing L1-Markowitz with the average performance of the naive k-portfolios, we find that neither is *a clear winner in out-of-sample performance*. The L1-Markowitz portfolios can consist of substantially more than $k = 15$ assets, and the might require the investor to take short positions, while the naive k-portfolios have a very simple structure. This suggests that the naive k-portfolios found by progressive resampling might be reasonable and simple alternatives to those found by the (approximately solved) L1-Regularised models. However, in 2004, 2009 and 2010, progressive resampling failed to find a portfolio with exactly 15 assets and the given constraints on return and risk. This is a downside of the approach that can be mitigated either by relaxing the constraints or allowing a larger portfolio (by setting k to a larger value).

5   Discussion, Limitations, and Conclusions

5.1   Related Approaches

We have discussed related work also earlier in context. Here we highlight some related research in more detail.

Randomised Portfolio Selection. The authors of [19] consider a problem very similar to our naive k-portfolio search task in the mean-variance framework.

However, their algorithm is based on randomised rounding of the solution to a semidefinite program, and is also not explicitly controlling for portfolio size as we do. Also, as discussed below, our problem formulation and solution approach are general enough to cover a variety of portfolio "quality measures", and can thus be applied also outside the mean-variance framework.

Pattern Sampling. E.g. frequent patterns [3,4] or subgroups [17] can efficiently be mined by biasing the sampling distribution to some extent in the same vein as is done in our algorithm. However, unlike work on pattern sampling, we are not interested in a representative sample that captures global properties of the entire pattern distribution. We simply want to find one portfolio that satisfies the constraints μ^* and σ^*.

Skyline Algorithms for Subsets of Size k. There exist efficient algorithms for computing the set of Pareto optimal subsets of size k [13,14]. (E.g. in database literature this set is called the "skyline" of a set of points.) These algorithms could, possibly with minor modifications, be used also to find good naive k-portfolios. However, they rely on linearity and submodularity of the objective functions, which may not apply for some risk/return measures. Progressive resampling should yield good solutions also in cases where the techniques from [13,14] are not applicable.

Genetic/Evolutionary Algorithms. Our approach could also be understood as a special kind of genetic algorithm with a binary fitness function: either an individual survives or it does not, without any intermediary outcomes. The main difference to textbook-style evolutionary optimisation methods is that progressive resampling obtains the next generation not via cross-overs and mutations of the current generation, but from a sampling distribution that characterises some global properties of the "survivors" from the previous generation.

Multiplicative Weights Update Method. The multiplicative weights update method [1] is a meta-algorithm that unifies a number of different algorithms e.g. from machine learning, flow problems, game theory, etc. It seems that the progressive resampling algorithm could also be understood as a variant of this approach. Exploring the connections between these methods in detail is left as future work, however.

5.2   Generalising the Naive k-portfolio Problem

Problem 1 is only a special case of a more general combinatorial problem:

Problem 2. Given finite set U, integer k, functions $f_i : 2^U \rightarrow \mathbb{R}$ and thresholds $\theta_i \in \mathbb{R}$ for $i = 1, \ldots, h$, find $X \subset U$ st. $|X| = k$ and $f_i(X) \leq \theta_i$ for all i, or determine that no such X exists.

In Problem 1 $h = 2$, the functions f_i are return and risk, and the thresholds θ_i are μ^* and σ^* (the inequality for return must be multiplied by -1). A related problem is *integer linear programming* (ILP), in particular the variant where we simply must determine if a given ILP has feasible solutions. Clearly ILP-feasibility is a special case of Problem 1, where the functions f_i are linear. Since ILP-feasibility is known to be NP-complete (A6 MP1 in [11]), the decision version of Problem 2 (that simply asks if an X that satisfies all constraints exists) is NP-complete as well.

5.3   Limitations

Our work has some limitations:

We Do not Find "The Optimal Portfolio". By defining the problem as a search of any 1/k portfolio that satisfies the user-specified constraints, we clearly do not find a portfolio that is in a strict sense optimal. However, it is not at all obvious that such an optimal portfolio would have better out-of-sample performance than a portfolio that is only "reasonably good" in training data. We argue that by making this more explicit in the problem definition, we to some extent can steer the investors' thinking away from optimality, and towards more realistic notions of the risk/return trade-offs.

No Naive k-Portfolio Might Satisfy the Investors' Constraints. In contrast to the classical Markowitz model, where the user (usually) only sets the desired return, and risk is minimised given this constraint, in our approach the user must set a constraint both on return and risk. If no naive k-portfolio satisfies the investors constraints, the options are either to relax the constraints or increase k. Knowing that no portfolios that satisfy the constraints exist may also be useful information for an investor.

Setting the Parameters is Difficult. It is true that to make an educated choice about μ^* and σ^* the investor should be given "global" information about all possible portfolios. A risk-return diagram (Fig. 1) can be helpful here as it shows how risk and returns are related, and what levels of risk one should be prepared to take for a given level of expected return.

The Mean-Variance Framework is Outdated Anyway. Clearly one can think of more sophisticated (and possibly more useful) ways to quantify portfolio risk than return variance. However, our problem definition nor algorithm do not require the use of any particular risk/return metrics, nor is it even constrained to the realm of risks and returns! Basically it can be used in exactly the same manner with any other approach to quantify relevant properties of a portfolio. Of course in theory (as well as practice) it is important that the resulting constraints are convex in the space of all portfolios.

5.4   Conclusion and Future Work

We have presented the problem of finding naive k-portfolios in the mean-variance framework, proposed a randomised algorithm for finding these, and conducted two experiments to study their performance. We found that there exist naive k-portfolios that have (in training data) comparable performance to optimised

L1-regularised Markowitz portfolios, and that these portfolios also have similar out-of-sample performance. Overall, our experiments confirm the well-known result that a very naive investment portfolio that simply allocates a small fraction of capital to *all possible* assets can be very competitive. Therefore, in the absence of other constraints the best choice for a small investor is a mutual fund or ETF that allocates its capital over all possible assets according to some simple mechanism, e.g. market capitalisation. However, if portfolio sparsity is of importance, and short-selling is not possible, the naive k-portfolios seem like a reasonable alternative to L1-Regularisation.

The generalisation of the naive k-portfolio problem (Problem 2) and the progressive resampling algorithm seem like intersting avenues for future work. For example, showing that progressive resampling is a *polynomial Monte Carlo algorithm* [18] for some variants of Problem 2 seems like an interesting open question. Also other questions of complexity and performance guarantees of Algorithm 1 are left for future studies.

Acknowledgements. This work has been supported by Academy of Finland (decisions 1308946 and 1314262).

References

1. Arora, S., Hazan, E., Kale, S.: The multiplicative weights update method: a meta-algorithm and applications. Theory Comput. **8**(1), 121–164 (2012)
2. Ban, G.Y., El Karoui, N., Lim, A.E.: Machine learning and portfolio optimization. Manag. Sci. **64**, 1136–1154 (2016)
3. Boley, M., Lucchese, C., Paurat, D., Gärtner, T.: Direct local pattern sampling by efficient two-step random procedures. In: ACM SIGKDD, pp. 582–590 (2011)
4. Boley, M., Moens, S., Gärtner, T.: Linear space direct pattern sampling using coupling from the past. In: ACM SIGKDD, pp. 69–77 (2012)
5. Brodie, J., Daubechies, I., De Mol, C., Giannone, D., Loris, I.: Sparse and stable Markowitz portfolios. PNAS **106**(30), 12267–12272 (2009)
6. DeMiguel, V., Garlappi, L., Nogales, F.J., Uppal, R.: A generalized approach to portfolio optimization: improving performance by constraining portfolio norms. Manag. Sci. **55**(5), 798–812 (2009)
7. DeMiguel, V., Garlappi, L., Uppal, R.: Optimal versus naive diversification: how inefficient is the 1/n portfolio strategy? Rev. Financ. Stud. **22**(5), 1915–1953 (2009)
8. Efron, B., Hastie, T., Johnstone, I., Tibshirani, R.: Least angle regression. Ann. Stat. **32**(2), 407–499 (2004)
9. Fastrich, B., Paterlini, S., Winker, P.: Constructing optimal sparse portfolios using regularization methods. Comput. Manag. Sci. **12**(3), 417–434 (2015)
10. Gao, J., Li, D.: Optimal cardinality constrained portfolio selection. Oper. Res. **61**(3), 745–761 (2013)
11. Garey, M.R., Johnson, D.S.: Computers and Intractability. WH Freeman, New York (1979)
12. Jagannathan, R., Ma, T.: Risk reduction in large portfolios: why imposing the wrong constraint helps. J. Financ. **58**, 1651–1684 (2003)
13. van Leeuwen, M., Ukkonen, A.: Discovering skylines of subgroup sets. In: ECMLP-KDD, pp. 272–287 (2013)

14. van Leeuwen, M., Ukkonen, A.: Same bang, fewer bucks: efficient discovery of the cost-influence skyline. In: SIAM SDM, pp. 19–27 (2015)
15. Luenberger, D.G.: Investment Science. Oxford University Press Inc., New York (1998)
16. Markowitz, H.: Portfolio selection. J. Financ. **7**(1), 77–91 (1952)
17. Moens, S., Boley, M.: Instant exceptional model mining using weighted controlled pattern sampling. In: IDA, pp. 203–214 (2014)
18. Papadimitriou, C.H.: Computational Complexity. Wiley, Hoboken (2003)
19. Xie, J., He, S., Zhang, S.: Randomized portfolio selection with constraints. Pac. J. Optim. **4**(1), 89–112 (2008)
20. Yen, Y.M., Yen, T.J.: Solving norm constrained portfolio optimization via coordinate-wise descent algorithms. Comput. Stat. Data Anal. **76**(C), 737–759 (2014)
21. Zheng, X., Sun, X., Li, D.: Improving the performance of MIQP solvers for quadratic programs with cardinality and minimum threshold constraints: a semidefinite program approach. INFORMS J. Comput. **26**(4), 690–703 (2014)

ICIE 1.0: A Novel Tool for Interactive Contextual Interaction Explanations

Simon B. van der Zon[1][✉], Wouter Duivesteijn[1][✉], Werner van Ipenburg[2][✉], Jan Veldsink[2][✉], and Mykola Pechenizkiy[1][✉]

[1] Eindhoven University of Technology, Eindhoven, The Netherlands
{s.b.v.d.zon,w.duivesteijn,m.pechenizkiy}@tue.nl
[2] Coöperatieve Rabobank U.A., Utrecht, The Netherlands
{werner.van.ipenburg,jan.veldsink}@rabobank.nl

Abstract. With the rise of new laws around privacy and awareness, explanation of automated decision making becomes increasingly important. Nowadays, machine learning models are used to aid experts in domains such as banking and insurance to find suspicious transactions, approve loans and credit card applications. Companies using such systems have to be able to provide the rationale behind their decisions; blindly relying on the trained model is not sufficient. There are currently a number of methods that provide insights in models and their decisions, but often they are either good at showing global or local behavior. Global behavior is often too complex to visualize or comprehend, so approximations are shown, and visualizing local behavior is often misleading as it is difficult to define what local exactly means (i.e. our methods don't "know" how easily a feature-value can be changed; which ones are flexible, and which ones are static). We introduce the *ICIE* framework (Interactive Contextual Interaction Explanations) which enables users to view explanations of individual instances under different *contexts*. We will see that various contexts for the same case lead to different explanations, revealing different feature interactions.

Keywords: Explanations · Feature contributions ·
Feature interactions · Model transparency · Awareness · Trust ·
Responsible analytics

1 Introduction

Within the domain of banking, black box models are used to predict fraud, money laundering and risk in lending, with their main advantage: speed. However, for real world application of such models, the outcome still has to be augmented by human experts for positive cases, as the bank has to be able to explain to its customers why a particular decision was made.

The standard accounting software packages of this World all provide features to perform what is colloquially known as a "What if? analysis": if we change the

© Springer Nature Switzerland AG 2019
C. Alzate et al. (Eds.): MIDAS 2018/PAP 2018, LNAI 11054, pp. 81–94, 2019.
https://doi.org/10.1007/978-3-030-13463-1_6

value of a cell, how does that affect other cells in my spreadsheet? It is a rather basic technique, and therefore by itself not particularly of interest for data miners. As a research field, we are typically more interested in reverse engineering, a "What happened? analysis". A classifier predicts a specific outcome, but which input attributes contribute most? Which input attributes make it happen? How far must we go in changing the input, in order to change the outcome?

In this paper, we propose to stack the two forms of analysis, to provide a more in-depth analysis of feature interaction and how it affects prediction. We do so by building onto the concept of SHAP contribution values [11]. Introduced in 1953 in the context of cooperative game theory, the original Shapley values [10] provide an answer to the question of how much of the total reward should be awarded to each member of a winning coalition, based on the individual contributions. This concept can be exploited [11] to determine to which degree a certain prediction outcome can be attributed to each individual input attribute.

However, not all real-world interactions can naturally be decomposed into individual contributions. For instance, a family history of clubfoot and maternal smoking individually both have a positive influence on the probability that the offspring displays isolated clubfoot. However, if the family history displays clubfoot, this has a multiplicative effect on the influence of smoking on the probability of clubfoot in the offspring [5]. Any decomposition of the final effect into an additive contribution to single input attributes will necessarily misrepresent what is really going on in this dataset. This is a fundamental problem, that this paper will also not solve.

Instead, our main contribution allows an end user to explore more convoluted interactions. A recent paper [6] introduced SHAP interaction values, which allow a user to find pairs of features which interact differently from their expected additive contributions. Although these values provide correct explanations of the case under observation, they are often still hard to interpret, and only give a limited view of what is actually happening (i.e. in the general context, where each feature is equally important). Instead, we propose to explore wider interactions, in the *ICIE* framework (Interactive Contextual Interaction Explanations). Under this framework, users can test various *contexts* to find explanations featuring unusual attribute interactions (i.e. *contexts* under which the "contextual SHAP values" change). Hence, ICIE enables users to find contexts (the "what if?" part) under which attributes interact unusually with the prediction (the "what happened?" part). Ultimately, this enables analysts to perform a more targeted investigation when verifying positive alerts.

2   Related Work

Baehrens et al. [1] propose methods which visualize a limited number of dimensions w.r.t. the class label assigned by the model, for individual classification decisions. Goldstein et al. [4], have a similar approach and observe the average global prediction when modifying these features. The LIME method [8] aims at constructing interpretable models in the vicinity of the case under explanation, and is prone to how exactly this vicinity is defined. Most current works on explanation of individual predictions use some form of sensitivity analysis to determine the impact of a feature. The EXPLAIN method [9] by Robnik-Šikonja and Kononenko, measures the impact of a removing a feature by comparing the original prediction to the average predictions for all of the feature's possible values. This method however cannot cope with interacting features which cannot impact the classifier's outcome alone (e.g. when $f(x) = a_1 \lor a_2$, for $x = \{a_1 = \text{true}, a_2 = \text{true}\}$, both a_1 and a_2 need to change in order to see that either one of them has an impact on the outcome, consequently, both will be assigned zero contribution). Štrumbelj et al. address this problem with IME [12], by observing each of the 2^n feature subsets (in case of binary classification) and hence has an exponential time complexity. While this method is capable of showing contributions for interacting features, it is not feasible for use on datasets with many attributes and/or attribute values. Štrumbelj et al. follow this up with an approximation algorithm [11] using the Shapley values from cooperative game theory [10], which makes computation feasible for larger domains. Lundberg et al. discuss theory and several properties of these SHAP values [7], provide an algorithm for efficient computation for ensembles of trees, and provide a generalization of the SHAP contributions which enables them to measure interaction between two features [6]. Finally, Martens and Foster provide algorithmic approaches to find explanations for high dimensional document data, in the form of minimal sets of words that change the outcome of the classifier when removed from the document [13].

As noted by Lundberg et al., many current methods for interpreting individual machine learning model predictions fall into the class of *additive feature contribution methods* [7]. This class covers methods that explain a model's output as a sum of real values attributed to each input feature. Additive feature contribution methods have an explanation model g that is a linear function of binary variables: $g(z') = \theta_0 + \sum_{i=1}^{n} \theta_i z'_i$, where $z' \in \{0,1\}^n$, n is the number of input features, and $\theta_i \in \mathbb{R}$. The z'_i variables typically represent a feature being observed or unknown, and the θ_i's are the feature contribution values.

3   Preliminaries

Given a dataset Ω, which is a bag of N records $x \in \Omega$ of the form $x = (a_1, \ldots, a_n, \ell)$, where $\{a_1, \ldots, a_n\}$ are the input attributes of the dataset, taken from some collective domain $\mathcal{X}$, and $\ell \in \{\ell_{\text{pos}}, \ell_{\text{neg}}\}$ is the binary class label. A model $f(x)$ can be trained to predict the class label ℓ.

To explain a model's decision, SHAP values can be computed for each of the attributes. These values reveal the additive contributions to the model's outcome and hence the sum of these values approximate the predicted class label closely. It is important to note that these values reveal the local contributions for a particular instance, consequently, the contribution for some feature a_1 can be positive for an instance x, while it can be negative for an instance x' (if feature a_1 interacts with one of the changed attributes in x').

Definition 1 (Additive SHAP contribution). *Let* $\theta_i(x)$ *be the additive SHAP contribution value for the* i^{th} *feature of an instance* x, *i.e.*

$$\theta_i(x) = \sum_{S \subseteq \mathcal{X} \setminus \{i\}} \frac{|S|!(n - |S| - 1)!}{n!} [f_x(S \cup \{i\}) - f_x(S)],$$

where n *is the number of attributes in the feature space* $\mathcal{X}$, *and* $f_x(S)$ *is the soft classifier output for model* f, *conditioned only on the features present in the feature subset* S.

Note that *negative* SHAP values represent feature-values contributing to the *negative* class and *positive* SHAP values to the *positive* class. Exact computation of the SHAP values is not feasible, but they can be approximated by the sampling algorithm described by Štrumbelj et. al. [11, Algorithm 1].

4   The ICIE Method

Providing an overview of only the SHAP contributions gives a limited view of the model's logic. After all, changing one feature, could (potentially completely) change the original contributions. We propose a method to manually explore *contexts* under which the SHAP values are different, and allow users to gain more confidence in for example a fraud alert.

We introduce "context" as our key element of interaction with the user. Context can be defined as a set of constraints, to describe a subspace of the feature space. For now, we restrict ourselves to context that take the form of the well-known descriptions from pattern mining (Definition 2). An instance is either covered by a description, or not. And hence can be used as a natural way to restrict the calculation of the SHAP values to a subspace of the feature space.

Definition 2 (Description). *A description is a set of constraints, mapping an instance from a domain, to a binary value:* $D = x \rightarrow \{true, false\}$.

SHAP values reflect the contribution of a feature a_i for an instance x and its predicted label ℓ, and are intuitively computed by taking the average change in soft prediction output for perturbed versions of x. For each perturbation the difference in output between a version with a_i, and one without a_i is summed. We use the context to restrict this perturbation space. Let this be clarified by an illustrative example. Suppose a model is trained on a dataset about car occasions to predict whether the price will be 'low' or 'high', and we are interested in the model's opinion on an instance x. Let $x = \{a_{\texttt{mileage}} = 250.000, a_{\texttt{fuel}} = $ 'gasoline'$\}$, where we are interested in the contribution of $a_{\texttt{mileage}}$. Suppose the model predicts $\ell_{\texttt{price}} = $ 'low', and feature $a_{\texttt{mileage}}$ with SHAP value $\theta_{\texttt{mileage}} = 0.5$ is the largest contributor for this decision. When the same case is now observed under the context of only `mileage` ≥ 200.000 cars. We likely observe that the contribution of `fuel` increases, as 'diesel' cars can handle a higher mileage, hence making the fuel type a more important selection criterion. Such interactions are not revealed by the SHAP values. In this work we propose a framework to let an analyst explore such scenarios.

4.1   Calculation of Contextual SHAP Values

We calculate SHAP values under a given context similarly to [11, Algorithm 1], with a twist. Instead of selecting a sample at random using $\pi(n)$ to perturb features (the set of all possible feature permutations), we select them from $\pi_D(n)$ (the set of all feature permutations satisfying the context D). The effect of the features in the context is amplified by disregarding the feature effects captured by the complement of the context. Formally $\pi_D(n)$ is defined as the set of tuples of the form $(d_1, \ldots, d_e, d_1^c, \ldots, d_{n-e}^c) \in \mathcal{X}$, where $A_D = \{d_1, \ldots, d_e\}$ is the set of attribute indices mentioned in D, and $(d_1^c, \ldots, d_{n-e}^c) \in d^c$ is the set of all random permutations of the complement of A_D. We refer to a SHAP value θ under context D by the notation θ^D. Note that under the general context, ICIE values reduce to the regular SHAP values.

4.2   UI for Context Exploration

Figure 1 shows the user interface of our application. The next sections discuss its various components. The software consists of two parts: (1) the computation of the classification models (from .csv data files) and (contextual) SHAP values are done on the server (Java); and (2) the user interface runs in the client browser (ECMAScript 6), and sends requests to the server via a MySQL database, where asynchronous Java workers are waiting to answer requests (i.e. computing (contextual) SHAP values).

SHAP Parameters Controller. In the top of the screen the parameters controller is shown. A dataset (with corresponding model) can be selected here. An instance from this dataset can be retrieved by inserting its **x.id**. The m corresponds to the SHAP value sampling criteria (from [11]). The first m value is the minimum number of samples drawn for each feature to get an initial estimation for the SHAP value of each feature, the second m number is the maximum number of samples that can be draw (multiplied by the number of features), and is divided based on the expect reduction in variance for the SHAP values. The button **fast/quality** can be used to set these two values to preset values.

Context Controller. Allows the user to manipulate the context. We restrict ourselves to contexts that describe subgroups of the data by using simple operators on a subset of features. In Fig. 1 a context representing non-Asian people, younger than 46 with "some" capital gain and education are represented. For ease of use, the **0/1** button next to the close button can toggle the context on/off.

Feature Contributions View. In this view, we present the contextual SHAP feature contributions. The user can manipulate the case under investigation here, allowing him/her to observe explanations for variations of the case. With each step in either changing the instance or the context, we highlight the aspects that

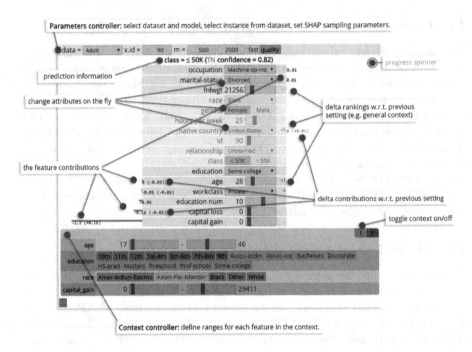

Fig. 1. The ICIE user interface. In the top of the screen, in the *parameters controller*, basic domain- and SHAP sampling parameters can be set. In the center of the screen, the feature importances for the current case (based on **x.id**) are shown (red bars on the left side for negative contributors, and green bars on the right for positive contributors). A user can modify the instance on the fly by changing its attributes. In the bottom of the screen the context is shown and can be modified. For each change that the user makes (either in context or the instance), the contribution values are recalculated, and the changes are reflected by the delta contributions (next to each contribution) and the delta rankings (next to the value selector for each attribute). (Color figure online)

change. The difference in contribution value is shown along with the change in ranking of the feature (based on the sorting by contribution).

5   Use Cases

We demonstrate our method on six datasets, all taken from the UCI repository [3]. This sample features a mixture of datasets having only binary/nominal or only numeric attributes, as well as datasets mixing attribute types. Characteristics of the used datasets, including statistics on the available attributes per type, can be found in Table 1. For each dataset Ω, a classifier is trained on a random sample of 80% of the data and some common quality measures are reported on the remaining 20% of the data. The models are depth-16 decision trees [2] (characteristics can be found in Table 2). Note that our approach is model agnostic, and we simply use decision trees for practical purposes.

For each dataset we show the additive contributions under the general context, along with a more specific context revealing interacting features (not possible to obtain from the general SHAP values). We comment on the found interactions revealed by the context and explain the process of navigating to the specific context. In the next section, some generalizable remarks on the "navigation process" are discussed.

Table 1. Dataset characteristics

Dataset	N	n	n_{bin}	n_{nom}	n_{num}
Ω_1 = Adult	48842	14	1	7	6
Ω_2 = Credit-card-default	3000	24	0	4	20
Ω_3 = German-credit	1000	21	1	13	7
Ω_4 = Mushroom	8124	22	6	16	0
Ω_5 = Tic-Tac-Toe	958	9	0	9	0
Ω_6 = Wisconsin	699	9	0	0	9

Table 2. Model characteristics

Dataset	Majority	Accuracy	Kappa	Precision	Recall	F_1-score
Ω_1 = Adult	0.92	0.83	0.45	0.91	0.91	0.91
Ω_2 = Credit-card-default	0.70	0.78	0.45	0.83	0.83	0.83
Ω_3 = German-credit	0.54	0.69	0.48	0.82	0.82	0.82
Ω_4 = Mushroom	0.53	1.00	1.00	1.00	1.00	1.00
Ω_5 = Tic-Tac-Toe	0.63	0.85	0.69	0.82	0.82	0.82
Ω_6 = Wisconsin	0.64	0.97	0.93	0.98	0.98	0.98

5.1   Adult

This dataset records instances on the annual revenue of individuals and discretizes the individuals in two categories (more than 50k, or less than or equal 50k). For this exploration example we observe the instance with $id = 90$ (the first instance has $id = 0$) from the Adult dataset. Figure 2a shows the "before" situation, and Fig. 2b shows the "after" situation (where the context is limited to `capital_gain=0`). In the before situation, it is clear which attribute is dominating the decision (namely `capital_gain`, with $\theta_{capital\_gain=0} = -1.48$). For this reason, we choose to restrict the context to this particular attribute; the intuition is that now the contributions will be computed only against instances with this same `capital_gain`, hence amplifying the inner effects in that subspace of the data. An unexpected finding reflected by our visualization is that in the general context `age` is the biggest positive contributor (with

$\theta_{\text{age}=28} = 0.02$), whereas in the specific context, age becomes a negative contributor (with $\theta_{\text{age}=28}^{\text{capital\_gain}=0} = -0.02$). This implies an interaction between the two features and can intuitively be interpreted as: generally age $= 28$ has a positive impact on the classification, but within the subspace of people with capital_gain $= 0$, this particular age, contributes negatively.

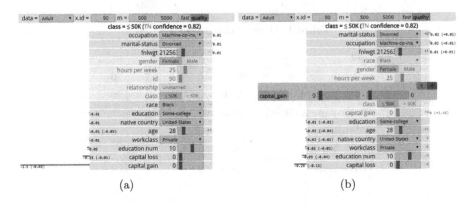

(a)   (b)

Fig. 2. Explanations for case 90 from the Adult dataset (we start counting from 0). On the left (Fig. 2a), we see the explanation in the general context, and on the right (Fig. 2b), we see the same explanation, but now under the context {capital_gain $= 0$}.

Another interesting observation for case 90 of the Adult dataset is presented in Fig. 3. The age attribute contributes differently depending on the particular age ranges we inspect. We observe that the positive contribution of age is amplified by a factor 10 in the context age ≤ 28 (Fig. 3a). Intuitively interpreted as: in the subspace of people with an age up to 28, age contributes substantially

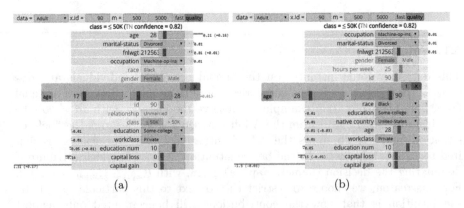

(a)   (b)

Fig. 3. Explanations for case 90 from the Adult dataset. On the left (Fig. 3a), we see the explanation under the context {age ≤ 28}, and on the right (Fig. 2b), we see the same explanation, but now under the context {age ≥ 28}.

more positively than in the general context (this makes sense as people who are 28 years old in the age group of people of age up to 28, generally make more money than younger ones). When we observe the complement of this context (age $\geq$ 28), the contribution of age is actually inverted (which makes sense for analogous reasons). The general context averages the contributions over the entire age range, where the negative contributions dominate this example, thus sketching a limited view which may be interpreted incorrectly.

5.2   Credit-Card-Default

This dataset records credit card clients in Taiwan from April 2005 to September 2005 and are divided in two groups: people who pay duly or people who default on their credit card payments. The model that was trained on this dataset reaches 89% accuracy. Figure 4 presents a client that defaulted on her credit card payment, with biggest contributor payment-amount-6 = 0 (the amount that was payed back six months ago). Note that the contribution of education = 'university' has a slight positive impact on defaulting (maybe because this person is actually still in university judging by the age). When inspected under context {payment-amount-6 = 0}, two interesting observations can be made. Firstly, within the attribute university grows by a factor 3, implying that the within this specific subgroup people in university are apparently more likely to default. Secondly, the importance of the age feature (towards paying duly) gets smaller, meaning that in this subgroup the positive effect of age is less.

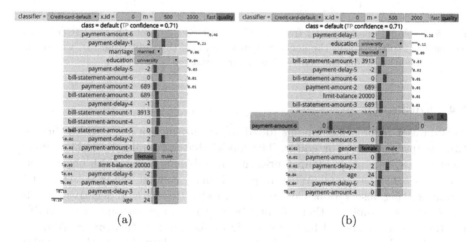

(a)   (b)

Fig. 4. Explanations for case 0 from the Credit-card-default dataset. On the left (Fig. 4a), we see the explanation in the general context, and on the right (Fig. 4b), we see the same explanation under the context {payment-amount-6 = 0}.

5.3   German-Credit

This dataset records clients from a German bank requesting a loan with a specified amount, purpose and duration for the loan. The clients are divided in two
groups: clients who did not manage to pay according to agreement (bad class)
and clients who did (good credit class). The model that was trained on this
dataset reaches 69% accuracy. Figure 5 presents a client that was assigned the
bad credit class, with biggest (bad-class) contributor `account-duration` = 48
(the duration of the loan was 48 months), and as biggest (good-class) contributor
`credit-amount` = 5951, which is apparently considered low by the model. If we
want to amplify the effects more, we restrict the context to a sub-range of the
biggest contributor, namely {`account-duration` ≥ 48} (only loans of at least
48 months). We can now observe that under this context the importance for
the biggest good-class contributor completely disappears, implying that while in
the general setting having this low credit amount is not a risk, in the context
of clients with a loan lasting 48 months or longer there actually is an increased
risk.

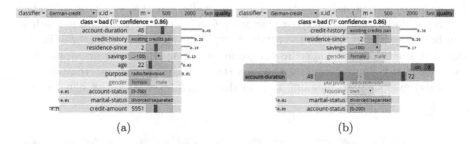

(a)   (b)

Fig. 5. Explanations for case 1 from the German-credit dataset. On the left (Fig. 5a),
we see the explanation in the general context, and on the right (Fig. 5b), we see the
same explanation under the context {`account-duration` ≥ 48}.

5.4   Mushroom

This dataset records mushrooms found in North-America and divides them in
two groups: poisonous or edible. The model that was trained on this dataset
reaches 99.8% accuracy. Figure 6 presents an edible mushroom with biggest contributor `odor` = 'Almond'. When inspected under context {`odor` = 'Almond'},
two interesting observations can be made. Firstly, there are no negative contributors anymore, implying that this attribute dictates the outcome (when verifying this claim, we indeed find that all 389 mushrooms with `odor` = 'Almond'
are edible). Note that this conclusion cannot be drawn from the contributions in the general context. Secondly, `stalk-root` = 'Bulbous' goes from
biggest negative contributor (with $\theta_{\texttt{stalk-root}='Bulbous'} = -0.14$) to biggest positive contributor (with $\theta^{\texttt{odor}='Almond'}_{\texttt{stalk-root}='Bulbous'} = 0.13$), telling us "yes" in general
`stalk-root` = 'Bulbous' contributes negatively, but when the mushroom smells
like almond, the opposite is true.

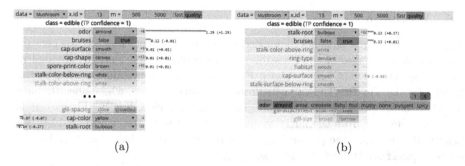

(a)   (b)

Fig. 6. Explanations for case 13 from the Mushroom dataset. On the left (Fig. 6a), we see the explanation in the general context, and on the right (Fig. 6b), we see the same explanation under the context {odor = 'Almond'}.

5.5 Tic-Tac-Toe

This dataset records all possible *end* games for the Tic-Tac-Toe game, with corresponding outcome (either × won, or × did not win; note that both a draw and ○ wins are counted as negatives). Figure 7 represents a game that was won by ×, with most important move center = ×. When observed in this particular context, we find that according to the model × always wins the game when it occupies the center (all contributions become 0, meaning that no other attribute influences the game). This is an important finding, as it points out one of the flaws of the model (we can think of many games where × occupies the center, but doesn't win the game).

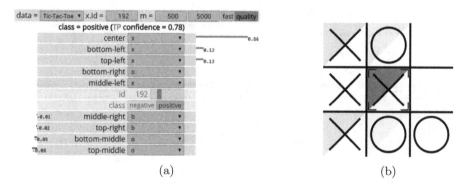

(a)   (b)

Fig. 7. Explanations for case 192 from the Tic-Tac-Toe dataset. For the sake of interpretability, we have drawn the board corresponding to this game. On the left (Fig. 7a), we see the explanation in the general context, and on the right (Fig. 7b), we see the same explanation under the context {center = ×}, this time visualized for more clarity. The top-left triangles in the cells of the board correspond to the contributions in the general context, and the ones in the bottom-right correspond to the contributions in the context, the (intensity of the) color corresponds to the contribution.

5.6   Wisconsin

This dataset records patients diagnosed with breast cancer. The records contain features about images of the (possible) tumor cells. The patients are divided in two groups (malignant and benign). First we report an instance similar to the finding for the Tic-Tac-Toe and Mushroom datasets, namely, where one attribute dictates the outcome. Figure 8 shows the same behavior, but in this case it is even more unclear from the initial visualization that `uniformity-of-cell-size` is actually dictating the prediction, which is reflected by observing it under the context {`uniformity-of-cell-size` ≥ 5}.

Next we use our method to inspect a wrong classification and try to find an explanation for the error. Figure 8 shows record 59, which is wrongly classified as 'benign'. In order to find out why the model made this mistake we start limiting the context to the biggest contributor for this case; `uniformity-of-cell-size`. To find out in which "direction" the positive contribution works, we observe the instance under two contexts: {`uniformity-of- cell-size` ≤ 3} and {`uniformity-of-cell-size` ≥ 3}. If the former removes the contribution of `uniformity-of-cell-size`, we conclude that this direction can't be used to alter the decision, else we argue the opposite. Figure 8d shows that the former holds, hence we modify `uniformity-of-cell-size` $= 3$ by replacing it with 4. We now see that negative is predicted (Fig. 8e). This may suggest that either

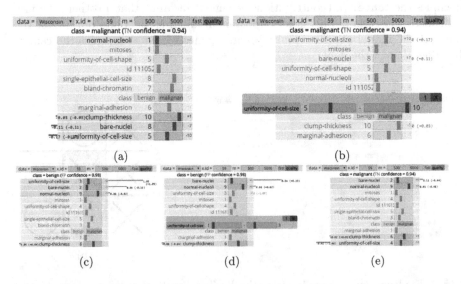

Fig. 8. Explanations for cases 50 and 59 from the Wisconsin dataset. On the top-left (Fig. 8a), we see the explanation in the general context, and on the top-right (Fig. 8b), we see the same explanation under the context {`uniformity-of-cell-size` $\geq$ 5}. On the bottom-left (Fig. 8d), we see the explanation in the general context, in the bottom-middle (Fig. 8d), we see the same explanation under the context {`uniformity-of-cell-size` ≤ 3}, and on the bottom-right (Fig. 8e), we see the instance where the value for `uniformity-of-cell-size` is replaced by 4.

the model is wrong here or a false measurement was recorded. Note that similar reasoning can be used in an automated search to find the least expensive path to changing the class label.

5.7   Guiding the Manual Search for Explanations

When investigating an explanation with our framework it is important to be aware of some basic strategies. It is often a good idea to start investigating the key-players (i.e. top-contributors) first, as it is less likely that contexts consisting of zero-contributors influence the other contributions (however not impossible, e.g. in the case where a zero-contributor is an average of an equal amount of positive contributors as negative contributors). In order to amplify the interactions that are happening *within* the top-contributors, one can set the context equal to its value; restricting the calculation of the SHAP values to this particular feature subspace. For features with large domains (especially numeric feature), the plain SHAP values provide a very limited amount of information, for example, when $\theta_{age=30} = 0.5$, we don't know whether it is increasing or decreasing over the interval of $[17–90]$ (or another interval), or whether its adjacent value $\theta_{age=31}$ would be substantially different or not. By using smartly positioned contexts (usually at these boundaries), we can make deductions from the resulting observations.

6   Discussion

Our approach shows to reveal novel information that cannot be obtained by observing the additive SHAP values alone. In particular it can be used to make local feature interactions visible by restricting the context to a dominating feature; to find interacting features (e.g. where one contributes positively in the general context, but negatively in a more specific context), which is of great value when it comes to understanding the classifier and the domain (for accurate classifiers); to help in inspecting wrong predictions of a classifier; and to get insights in the "direction" of the feature contributions when it comes to numeric features.

We are currently exploring strategies to automatically discover interesting contexts, making the manual search less time consuming, and allowing us to find contexts consisting of more features. Ultimately, such tooling helps in answering more involved questions, such as: "given a context, what is the possible/likely adversarial activity leading to a change in this predicted label?" Helping experts in financial domains to target the right sources of information when verifying an alert.

References

1. Baehrens, D., Schroeter, T., Harmeling, S., Kawanabe, M., Hansen, K., Müller, K.-R.: How to explain individual classification decisions. J. Mach. Learn. Res. **11**(Jun), 1803–1831 (2010)
2. Breiman, L., Friedman, J.H., Olshen, R.A., Stone, C.J.: Classification and Regression Trees. CRC Press, New York (1989)
3. Dheeru, D., Karra Taniskidou, E.: UCI machine learning repository (2017). http:// archive.ics.uci.edu/ml
4. Goldstein, A., Kapelner, A., Bleich, J., Pitkin, E.: Peeking inside the black box: visualizing statistical learning with plots of individual conditional expectation. J. Comput. Graph. Stat. **24**(1), 44–65 (2015)
5. Honein, M.A., Paulozzi, L.J., Moore, C.A.: Family history, maternal smoking, and clubfoot: an indication of a gene-environment interaction. Am. J. Epidemiol. **152**, 658–665 (2000)
6. Lundberg, S.M., Erion, G.G., Lee, S.-I.: Consistent individualized feature attribution for tree ensembles, arXiv preprint arXiv:1802.03888 (2018)
7. Lundberg, S.M., Lee, S.-I.: A unified approach to interpreting model predictions. In: Conference Proceedings on Advances in Neural Information Processing Systems, pp. 4768–4777 (2017)
8. Ribeiro, M.T., Singh, S., Guestrin, C.: Why should i trust you? Explaining the predictions of any classifier. In: Proceedings of KDD, pp. 1135–1144 (2016)
9. Robnik-Šikonja, M., Kononenko, I.: Explaining classifications for individual instances. IEEE Trans. Knowl. Data Eng. **20**(5), 589–600 (2008)
10. Shapley, L.S.: A value for n-person games. Contrib. Theory Games **2**(28), 307–317 (1953)
11. Štrumbelj, E., Kononenko, I.: Explaining prediction models and individual predictions with feature contributions. Knowl. Inf. Syst. **41**(3), 647–665 (2014)
12. Štrumbelj, E., Kononenko, I., Robnik-Šikonja, M.: Explaining instance classifications with interactions of subsets of feature values. Data Knowl. Eng. **68**(10), 886–904 (2009)
13. Martens, D., Foster, P.: Explaining data-driven document classifications. MIS Q. **38**(1) (2014)

Testing for Self-excitation in Financial Events: A Bayesian Approach

Ali Caner Türkmen$^{(\boxtimes)}$ and Ali Taylan Cemgil

Department of Computer Engineering, Boğaziçi University, Bebek,
34342 Istanbul, Turkey
{caner.turkmen,taylan.cemgil}@boun.edu.tr

Abstract. Self-exciting temporal point processes are used to model a variety of financial event data including order flows, trades, and news. In this work, we take a Bayesian approach to inference and model comparison in self-exciting processes. We discuss strategies to compute marginal likelihood estimates for the univariate Hawkes process, and describe a Bayesian model comparison scheme. We demonstrate on currency, cryptocurrency and equity limit order book data that the test captures excitatory dynamics.

1 Introduction

Many real-world data mining applications, including those in finance, entail modeling event occurrences in a continuous time setting. Examples of such data abound in finance; including order flows [3], trades [1], news [12], price jumps, volatility spikes, etc. Temporal point processes, statistical models of points scattered along the real line, are often the primary models used to address these data sets.

The Poisson process (PP) is one such statistical model that assumes independence among occurrences. Points are assumed to occur without any interaction, sometimes described as *completely randomly* [6]. PPs have been used in finance for modeling discrete event systems, e.g. limit orders [3]. While PPs lead to convenient mathematics for computing many quantities of interest analytically, they fail our simple intuition that financial events are seldom independent of one another, i.e. that they *excite* each other.

Self-exciting point processes, specifically Hawkes processes (HP) [7], are recently growing more common in quantitative finance [2] as well as machine learning literatures [8,9]. First explored in the backdrop of seismology, HPs assume causal, linear non-negative excitation behavior among occurrences. This is why they have been considered especially suited to modeling financial discrete events.

Typically, HPs are applied towards prediction tasks. Maximum likelihood estimates of model parameters are fit to an observation, a collection of occurrence timestamps, that are assumed to arise from the process. Model validation or

C. Alzate et al. (Eds.): MIDAS 2018/PAP 2018, LNAI 11054, pp. 95–102, 2019.
https://doi.org/10.1007/978-3-030-13463-1_7

selection is then performed through predictive likelihood, or some other cross-validation metric, used to determine how good the fit is on a held out sample. Here, instead, we present a method of model selection (or equivalently, hypothesis testing) for self-exciting point process models. We take a Bayesian approach, and describe approximate inference and marginal likelihood estimation schemes. We present preliminary experiments on high frequency currency, cryptocurrency and equity limit order book data. Among a family of Bayesian inference methods, we posit that Laplace approximation to model evidence is best suited to the problem at hand.

In Sect. 2 we first give a brief overview of self-exciting processes and Bayesian model selection before describing our inference scheme. In Sect. 3, we present a set of preliminary findings on currency price, equity order book, and cryptocurrency event sets, before concluding in Sect. 4.

2   Model

2.1   Hawkes Process

Let $\{N(t)\}_{t\in\mathbb{R}_+}$ denote a counting process, a jump process where jump sizes are $+1$ and $N(0) = 0$. Furthermore, we will use the overloaded notation $N(a, b]$ to refer to the number of jumps (or equivalently, *points*) in the interval $(a, b]$ – also a random variable. In correspondence to a temporal point process, we think of $N(t)$ as the number of points –event occurrences such as orders or transactions– until time t.

Homogeneous Poisson processes are characterized by complete independence and stationarity assumptions. We have that $N(a, b]$ and $N(c, d]$ are independent random variables given that $(a, b]$ and $(c, d]$ are disjoint intervals on the real line. Furthermore, by stationarity we have that $\langle N(a, b]\rangle = \langle N(a + \tau, b + \tau]\rangle$ for all τ, where we let $\langle.\rangle$ denote the expectation operator. However, it is these two assumptions that limit a realistic modeling of sequences of events that might as well have influenced each other.

Working with general classes of point processes where point occurrences are interdependent is difficult – both theoretically and computationally [6]. One alternative that leads to both mathematical and computational convenience is a class of temporal point processes (or, equivalently, counting processes), determined by a *conditional intensity* function [6]. Concretely, let λ^* denote the conditional intensity function of a self-exciting point process[1], defined by

$$\lambda^*(t) \triangleq \lim_{\delta\downarrow 0} \delta^{-1}\langle N(t, t + \delta]|\mathcal{H}_t\rangle.$$

Here we use $\mathcal{H}_t$ to denote the *history* of events up to time t[2]. Note that setting $\lambda^*(t) = \nu(t)$, a deterministic measurable function of t, would simply yield a (nonhomogeneous) Poisson process.

---

[1] We follow the notation λ^* of [6], where the superscript $*$ serves as a reminder that the intensity function is dependent on the history up to time t, $\mathcal{H}_t$.

[2] Formally, $\mathcal{H}_t$ can be seen as the natural filtration, an increasing sequence of σ-algebras, with respect to which we define the conditional expectation operator.

HPs arise as one of the simplest examples of point processes defined through a conditional intensity [4,6]. They model *linear self-excitation* behavior, where the instantaneous probability of an event occurrence is given by a linear combination of the effects of past events. A (univariate) HP is a point process determined by the conditional intensity function [6,7].

$$\lambda^*(t) = \mu + \sum_{t_j < t} \varphi(t - t_j). \tag{1}$$

Here $\mu > 0$ is the constant background (exogenous) intensity function. $\varphi : \mathbb{R}_+ \to \mathbb{R}_+$ is the *triggering kernel*, an often monotonically decreasing function that governs self-excitation.

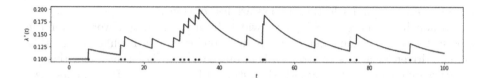

Fig. 1. Intensity function of a Hawkes process with exponential delay density

We will be concerned with the case $\varphi(x) = \alpha\theta \exp(-\theta x)$, where $\alpha \in [0,1), \theta > 0$. Here since $\int \theta \exp(-\theta x)d\theta = 1$, we can interpret the triggering kernel in terms of its parameters. α governs the *infectivity* or the average number of new events that are *triggered* by an event. The remaining part is the exponential density for the length of the delay between events triggering each other. Note that $\alpha < 1$ is required for stationarity.

One can think of the intensity as a stochastic process itself, which is *excited* every time a jump occurs stochastically on the underlying process $N(t)$. That is, a jump in $N(t)$ leads to a jump of size α in λ^*. This effect then decays according to a schedule determined by the decay factor in φ, which in the case above, was taken as an *exponential* decay proportional to $\exp(-\theta\Delta t)$. We illustrate this effect in Fig. 1.

We refer the reader to the review by Bacry et al. [2] for further details on HP and their varied applications in quantitative finance.

Finally, let us note that for any conditional intensity point process the likelihood of finitely many points $\Pi = \{t_i\}_{i=1}^N$ where $0 < t_1 < \cdots < t_N < T$ on a bounded interval $(0,T]$ is given by

$$p(\Pi|\lambda^*) = \exp\left(-\int_0^T \lambda^*(s)ds\right) \prod_i^N \lambda^*(t_i),$$

where the conditional intensity function $\lambda^*(x)$ uniquely determines the process. For Poisson processes, granted that the *compensator* $-\int_0^T \lambda(s)ds$ can be computed, the evaluation of the likelihood is trivial. This is not the case in general, however. Note that the computation of the likelihood for a general HP defined as in (1) would take time $O(N^2)$, as each intensity evaluation takes time linear in the number of events. This crucial aspect prohibits the use of likelihood-based inference, including many Bayesian methods, in general. In the exponential kernel HP case, however, both the log likelihood and its gradient can be computed in linear time owing to the *memoryless* property. In the sequel, we constrain our attention to HP parameterized as such.

2.2   Bayesian Model Comparison

As mentioned previously, point processes are used mainly as models of discrete events occurring asynchronously in continuous time. Compared to discrete-time models that are often used in econometrics or time series forecasting, the methods of comparing and selecting models are less obvious.

Although HPs have been explored widely in finance, existing works often use cross-validation – basing model comparison on predictive likelihood, or other domain-driven measures of error on held out data. On the other hand, there is earlier work on frequentist hypothesis testing of HP vs PP [5]. In this paper, we present work in progress regarding a Bayesian approach – bringing the advantages (and potential pitfalls) of encoding prior assumptions on model parameters and deriving intuitive tests of model validity.

In Bayesian model comparison, one judges models through *marginal (integrated) likelihoods*, using the same calculus of probability that one judges parameter configurations of a fixed model. Let $p(\Pi|\Theta)$ denote the data likelihood, and $p(\Theta)$ a prior distribution under a certain model. Our aim is to compute the marginal likelihood

$$p(\Pi) = \int p(\Pi|\Theta)p(\Theta)d\Theta,$$

where we let Θ denote the vector of all model parameters. Intuitively, this quantity can be read as $\langle p(\Pi|\Theta) \rangle_{p(\Theta)}$, i.e. the expected likelihood that a given model will assign to data Π, as parameters are drawn from the prior $p(\Theta)$. Note that this quantity comes with "Occam's razor" included, i.e. high-dimensional models with diffuse priors are automatically penalized. One can then use the marginal likelihoods of two different models to compare them.

Let p_1, p_0 denote marginal likelihoods under two different models. The ratio

$$BF = \frac{p_1(\Pi)}{p_0(\Pi)} \tag{2}$$

is known as the *Bayes factor*. Bayesian hypothesis tests are performed by calculating the marginal likelihood under the null (p_0), as well as the alternative (p_1) hypotheses, and computing BF. $BF > 10$ is taken as strong evidence that the first model (p_1) better explains the observations. Similarly, many models (or prior configurations) can be compared on the same footing.

2.3 Proposed Method

Here we propose a simple hypothesis test for "self-excitation" behavior in financial events. We calculate the Bayes factor (2) by taking a homogeneous PP as the null hypothesis (p_0), and an exponential-decay HP as given in (1) as the alternative (p_1). In doing so, we explore methods of marginal likelihood estimation for HP, which also paves the way to comparing HP models.

We equip both models (p_0, p_1) with appropriate prior distributions. In the former, we choose a Gamma distribution for the constant intensity parameter. The Gamma distribution is conjugate to the PP likelihood, making marginal likelihood computation analytically tractable. For HP, parameters μ, α, θ are given Gamma, Beta and Gamma priors respectively.

Marginal likelihood for HPs is intractable under any choice of prior, and we must resort to an approximation. Yet, this approximation is still made difficult by computational challenges related to the likelihood, outlined above. For example, one sampling-based alternative for marginal likelihood estimation, annealed importance sampling [11], requires a large number of likelihood computations before a single weighted sample can be drawn. This prohibits a realistic application of this method for HPs with large observed samples.

However, especially in the high-frequency context, we can invoke another approximation method. Financial continuous time data sets, unlike earthquakes, are characterized by large sample sizes. We find that this leads to peaked, unimodal posteriors, with which we can turn to Laplace approximation to the marginal likelihood [10].

We approximate the posterior with a multivariate Gaussian distribution centered around the posterior mode, $\Theta^* = \arg\max p(\Theta|\Pi)$. Given the posterior potential $\varphi(\Theta) = p(\Theta|\Pi)p(\Theta)$, we approximate $p(\Pi) = \int d\Theta \varphi(\Theta)$ via

$$\ln p(\Pi) \approx \ln \varphi(\Theta^*) + \frac{3}{2} \ln 2\pi - \frac{1}{2} \ln |H|,$$

where $H = \nabla^2 - \varphi(\Theta^*)$ is the Hessian of $-\varphi$ evaluated at the mode.

This method reduces marginal likelihood estimation to a series of simple steps. First, maximum a posteriori (MAP) estimates of HP are obtained. This can be achieved via expectation maximization, as well as gradient-based methods in the simple case of univariate HP. The Hessian H can be approximated numerically or computed exactly. Software for estimating marginal likelihood, as well as other tasks such as posterior inference under univariate Bayesian HP, is made available online[3].

3 Experiments

Our experiments cover a range of financial event sets. FX are high frequency (millisecond range) tick events in an interbank currency exchange, previously

---

[3] http://www.github.com/canerturkmen/hawkeslib, and on the Python Package Index (PyPI) as hawkeslib.

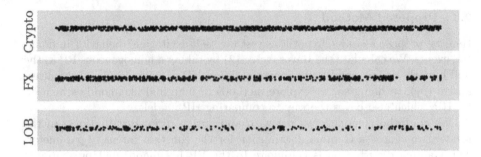

Fig. 2. Data samples from the three data sets. x-axis denotes time of occurrence, y-axis is random noise for better visibility

Table 1. Results of experiments on financial data sets. N denotes the number of occurrences in the data set. BF is the computed Bayes factor. Bayesian credible intervals at 95% are given for α, θ

Data Set	Asset	N	BF	α		θ	
				Lower	Upper	Lower	Upper
FX	GBPUSD	1000	$>10^8$	0.62	0.75	5.46	8.06
	EURUSD	1000	$>10^8$	0.53	0.64	6.26	8.37
	USDJPY	1000	$>10^8$	0.61	0.73	5.15	7.12
Crypto	USDT-ETH	5684	0.11	0.15	0.38	0	0.02
	BTC-XRP	5710	0.04	0.08	0.45	0	0.02
	BTC-ETH	5499	3.69	0.12	0.34	0	0.02
LOB	GARAN	1000	$>10^8$	0.88	0.99	8.8	18

investigated using HP [13]. We model three large-volume currency pairs selected at random. Crypto are price increase events on three large-cap cryptocurrencies on the cryptocurrency exchange Bittrex sampled at five minute (low) frequency. Finally LOB are limit order arrivals in a large-cap bank stock in the Turkish equity exchange, Borsa Istanbul, sampled at very high frequency (nanosecond range). Samples of each data set are given in Fig. 2. In FX and LOB , we limit event sets to 1000 events, roughly to 10 min of trading. Observe that in both data sets, the data cluster around certain points in time. This effect is less pronounced in Crypto .

We report the results of our tests, where we calculate the Bayes factor as described in Sect. 2.3. We further present 95% Bayesian credible intervals for the triggering kernel parameters, where we use simple random walk Metropolis (RWM) [10] algorithm to draw from the posterior.

We present the results in Table 1. The test accurately captures that low-frequency price jumps do not present sufficient evidence in favor of self-excitation. In FX and LOB , however, we find overwhelming evidence that HP outperforms PP. Note however that, if one were to register only large return jumps as events, HPs could still fit the data at lower frequencies. This is not surprising, its analogue in the discrete-time setting would be known as *volatility clustering*.

There are, however, two issues we must address. First, Bayesian analysis is well known to be sensitive to choice of priors. In our analyses, we find that large data sets easily mitigate this effect. In Fig. 2, we change the scale hyperparameter of the prior for θ, the delay distribution. We find that, except for unrealistic choices of priors which set the average delay to less than 0.01 ms, the conclusion is largely unaffected. Varying other hyperparameters lead to similar conclusions.

Finally, let us note that this paper and many others in the field assume constant background intensity μ. The test in this paper also assumes a homogeneous PP as the null hypothesis. However, the *exogenous* process that governs financial events is often not stationary. For example, financial events follow intraday, weekly and yearly cycles. Our test, and many other investigations in HP, are prone to capturing this effect and explaining it away using the *endogenous* component of HP. We test this effect using a toy data set drawn from a non-homogeneous PP with intensity $\lambda(t) \propto \exp(\sin t)$ (see, e.g. Figure 4). On this data, our test easily passes (rejects PP), although the nonstationarity is purely exogenous. In our experiments, we mitigate the potential effect of periodicity by sampling short time intervals (Fig. 3).

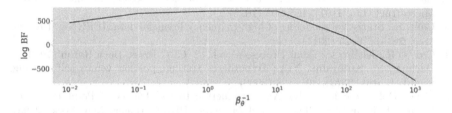

Fig. 3. Logarithm of the Bayes factor as the scale hyperparameter of $p(\theta)$ is changed; for EURUSD in the **FX** data set.

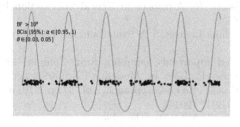

Fig. 4. A draw from a nonhomogeneous Poisson process with periodic intensity

4   Conclusion

We combined techniques from Bayesian machine learning and evolutionary point processes for modeling high-frequency financial data. We cast HP to a Bayesian setting, and discussed the computation of a Bayesian model comparison scheme for testing "self-excitation" behavior in financial events as well as posterior inference. Early experiments confirm basic intuition regarding high-frequency financial events.

Our method can be used to capture self-excitation effects in financial discrete event data, much in the same way conditional heteroskedasticity models capture volatility clustering. However, the test assumes that background intensities are stationary, and can lead to pitfalls in financial analysis. This issue constitutes the next step to this study.

Acknowledgement. We gratefully acknowledge the support of Scientific and Technological Research Council of Turkey (TUBITAK), under research grant 116E580.

References

1. Bacry, E., Dayri, K., Muzy, J.F.: Non-parametric kernel estimation for symmetric Hawkes processes. Application to high frequency financial data. Eur. Phys. J. B **85**(5), 157 (2012)
2. Bacry, E., Mastromatteo, I., Muzy, J.F.: Hawkes processes in finance. Mark. Microstruct. Liq. **1**(01), 1550005 (2015)
3. Cont, R.: Statistical modeling of high-frequency financial data. IEEE Sig. Process. Mag. **28**(5), 16–25 (2011)
4. Cox, D.R., Isham, V.: Point Processes, vol. 12. CRC Press, Boca Raton (1980)
5. Dachian, S., Kutoyants, Y.A.: Hypotheses testing: poisson versus self-exciting. Scand. J. Stat. **33**(2), 391–408 (2006)
6. Daley, D.J., Vere-Jones, D.: An Introduction to the Theory of Point Processes: Volume I: Elementary Theory and Methods. PIA. Springer, New York (2003). https://doi.org/10.1007/b97277
7. Hawkes, A.G.: Point spectra of some mutually exciting point processes. J. R. Stat. Soc. Ser. B (Methodol.) **33**, 438–443 (1971)
8. Linderman, S., Adams, R.: Discovering latent network structure in point process data. In: International Conference on Machine Learning, pp. 1413–1421 (2014)
9. Mei, H., Eisner, J.M.: The neural Hawkes process: a neurally self-modulating multivariate point process. In: Advances in Neural Information Processing Systems, pp. 6757–6767 (2017)
10. Murphy, K.P.: Machine Learning: A Probabilistic Perspective. MIT Press, Cambridge (2012)
11. Neal, R.M.: Annealed importance sampling. Stat. Comput. **11**(2), 125–139 (2001)
12. Rambaldi, M., Pennesi, P., Lillo, F.: Modeling FX market activity around macroeconomic news: a Hawkes process approach, vol. 67, p. 210. arxiv preprint. arXiv preprint arXiv:1405.6047 (2014)
13. Türkmen, A.C., Cemgil, A.T.: Modeling high-frequency price data with bounded-delay Hawkes processes. In: Corazza, M., Durbán, M., Grané, A., Perna, C., Sibillo, M. (eds.) Mathematical and Statistical Methods for Actuarial Sciences and Finance, pp. 507–511. Springer, Cham (2018). https://doi.org/10.1007/978-3-319-89824-7_90

A Web Crawling Environment to Support Financial Strategies and Trend Correlation

(Extended Abstract)

Giovanni Ponti[1]([✉]), Giuseppe Santomauro[1], Fiorenzo Ambrosino[1],
Giovanni Bracco[1], Antonio Colavincenzo[2], Matteo De Rosa[1], Agostino Funel[1],
Dante Giammattei[1], Guido Guarnieri[1], and Silvio Migliori[1]

[1] ENEA - ICT Division - Portici Research Center (NA), Portici, Italy
{giovanni.ponti,giuseppe.santomauro,fiorenzo.ambrosino,giovanni.bracco,
matteo.derosa,agostino.funel,dante.giammattei,guido.guarnieri,
silvio.migliori}@enea.it
[2] Accenture Technology Solutions s.r.l. - Assago (MI), Milan, Italy
antonio.colavincenzo@accenture.com

Abstract. We provide an overview on the development and the integration in ENEAGRID of a web crawling tool to retrieve data from the Web, manage and display it, and extract relevant information. We collected all these instruments in a collaborative environment called *Web Crawling Virtual Laboratory*, offering a GUI to operate remotely. Finally, we describe an ongoing activity on semantic crawling and data analysis to discover trends and correlations in finance.

Keywords: Web crawling · Big data · Machine learning ·
Market trends

1 Introduction

Internet is certainly the World's largest data source. Web data has characteristics that involve a considerable effort of analysis and organization. The ability of extracting strategical information in big data from the Web is becoming a crucial task that involves several contexts, such as cyber security, business intelligence, and finance. All the applications in these fields have to face with computational and storing issues. For this reason, the advanced computing center of ENEA Portici, hosting the ENEAGRID/CRESCO infrastructure [2] offers the possibility to perform this activity. In the following, we introduce the web crawling environment integrated in ENEAGRID to retrieve and analyze data from the Web, and we provide some details on a work-in-progress activity in finance describing how to obtain financial information and correlation with market trends.

© Springer Nature Switzerland AG 2019
C. Alzate et al. (Eds.): MIDAS 2018/PAP 2018, LNAI 11054, pp. 103–107, 2019.
https://doi.org/10.1007/978-3-030-13463-1_8

2   Web Crawling and Web Data Analysis in ENEAGRID

A crawling technique analyzes systematically and automatically the content of a network to search for documents to download. Web crawlers are based on a list of URLs to visit that is continuously updated by new records retrieved by parsing the explored web pages. In the next, we provide a description of our web crawling environment installed and configured in ENEAGRID.

2.1   Web Crawling Tool: *BUbiNG*

We resorted to *BUbiNG* [1] as the web crawling product to integrate in ENEAGRID. This software allows the parallel execution of multiple crawling agents. Each agent communicates with each other one to ensure not repeated visits of same pages and to balance the computational load. *BUbiNG* also allows to set up at runtime all configuration options in a single parameter setting file, such as thread number and initial seeds. *BUbiNG* saves contents in compressed *warc.gz* files. This data compression is very important because it allows to save space up to around 80%.

2.2   Virtual Laboratory and Web Application

We created a collaborative *Web Crawling Project* integrated in ENEAGRID. Here, the main issue consisted in harmonize the tool in a typical HPC environment to exploit infrastructure resources, that are computational nodes, networking, storage systems, and job scheduler. All the web crawling instruments are collected in a ENEAGRID virtual laboratory, named *Web Crawling*[1]. The virtual lab has a public web site (Fig. 1(a)) where information about the research

(a)   **(b)**

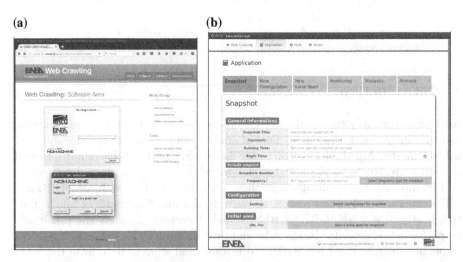

Fig. 1. (a) The virtual lab site. (b) The virtual lab GUI.

---

[1] http://www.afs.enea.it/project/webcrawl/.

activity is collected, and a web application (Fig. 1(b)) to submit snapshot and to use tools for analysis, displaying and clustering of web data.

2.3   Tests and Experimental Results

We performed experiments to analyze the performance of our solution for web crawling tool integrated in the ENEAGRID infrastructure. For this reason, we designed two types of experiments. In the first one, we performed long-time crawling sessions (of more than 8 h), in order to assess the ability of the tool in crawling and storing web contents at a high network speed, i.e., efficiency and robustness. The second experiment consisted in periodic crawling to test software reliability, a typical scenario to collect periodic snapshots to analyze changes in the network. Both tests provide good results [3].

3   Proposal of Current Development

We are currently working in extending our tool to support *semantic crawling* and apply it in finance, in order to discover how news and discussions in the Web on a specific topic are correlated with market trends and how can influence them.

3.1   Thematic Web Crawling

Working on proper crawling settings and pre-processing strategies, it is possible to have a reduced version of the crawled dataset on a specific topic. In this way we reached two main goals: saving memory space and speeding up the post-crawling indexing time. To obtain this result, we have developed a proper filter that selects web pages according to the topic. Such a filter does not take into account only at the page body, but also title and tags. We integrated it into BUbiNG source code (in *JAVA*) to have thematic snapshot sessions.

3.2   Web Crawling for Financial Strategies

By using the filtered dataset we aim to discover news and discussions in the Web on a specific topic. Information retrieval and deep learning algorithms can be employed to extract strategical information. More specifically, we want to reach two important results: (i) searching for a correlation index of web news with market trends and their influence, (ii) and developing a tool in order to predict a price behaviour and then to adopt appropriate trading strategy. Below we explain the five steps that we have considered:

1. First of all, for any day d_i we run a web crawling filtered on web news about a financial topic to build a dataset D_i of $g_{i,j}$ web pages:

$$D_i = \{g_{i,1}, g_{i,2}, \ldots, g_{i,N_i}\};$$

2. After, for any web page $g_{i,j}$ we apply a *Sentiment Analysis* algorithm, based on Natural Language Processing, (e.g. the VADER Sentiment Analysis coded in $JAVA^2$) to compute a weight of positive/negative opinion:

$$w_{i,j} = w(g_{i,j}) \in [-1; +1], \quad \forall j \in [1; N_i];$$

3. Then, we compute a normalized daily opinion index:

$$w_i = \frac{\sum_{j=1}^{N_i} w_{i,j}}{N_i};$$

4. By means of a machine learning approach, we train a neural network (i.e., a Recurrent Neural Network - RNN) to estimate the daily increasing/decreasing rate r_i for the asset:

$$r_i = \frac{p_{i+1} - p_i}{p_i},$$

where p_{i+1} is the estimated price at the day d_{i+1} obtained by the RNN computation.

5. Finally, we compute a correlation between rate R and opinion index W applying the *Pearson correlation coefficient*:

$$cov(R, W) = \frac{E[RW] - E[R]E[W]}{\sqrt{E[R^2] - E[R]^2}\sqrt{E[W^2] - E[W]^2}}.$$

For our purpose, in a day d_i, we want to discover a correlation between the expected increasing/decreasing rate r_i and the overall opinion index w_i.

4   Conclusions

To summarize, we provided a parallel implementation of a web crawling product to periodically download contents from web and to analyze them. The tool is fully integrated in our HPC ENEAGRID/CRESCO infrastructure, in order to use computation and storage power. Currently we are equipping our framework with a sentiment analysis tool and training a neural network to correlate opinions and price trend. In the future work we want to perform experiments to tune our framework and refine our semantic filter to obtain a more accurate dataset.

Acknowledgements. The computing resources and the related technical support used for this work have been provided by ENEAGRID/CRESCO High Performance Computing infrastructure and its staff [2]. ENEAGRID/CRESCO High Performance Computing infrastructure is funded by ENEA, the Italian National Agency for New Technologies, Energy and Sustainable Economic Development and by Italian and European research programmes, see http://www.cresco.enea.it/english for information.

---

[2] https://github.com/apanimesh061/VaderSentimentJava.

References

1. Boldi, P., Marino, A., Santini, M., Vigna, S.: BUbiNG: massive crawling for the masses. CoRR abs/1601.06919 (2016)
2. Ponti, G. et al.: The role of medium size facilities in the HPC ecosystem: the case of the new CRESCO4 cluster integrated in the ENEAGRID infrastructure, pp. 1030–1033 (2014)
3. Santomauro, G., et al.: A collaborative environment for web crawling and web data analysis in ENEAGRID. In: DATA 2017, 24–26 July 2017, Madrid, Spain, pp. 287–295 (2017)

PAP 2018: The 2nd International Workshop on Personal Analytics and Privacy

The 2nd International Workshop on Personal Analytics and Privacy (PAP 2018)

Workshop Description

In the era of Big Data, every single user of our hyper-connected world leaves behind a myriad of digital breadcrumbs while performing her daily activities. This enormous amount of *personal data* can be exploited to improve the lifestyle of each individual by extracting and analyzing user's behavioral patterns like the items frequently purchased, the routinary movements, the favorite sequence of songs listened, etc. Moreover, users have a very limited capability to control and exploit their personal data: although some user-centric models like the *Personal Information Management System* and the *Personal Data Store* are emerging, currently there is still a significant lack in terms of algorithms and models specifically designed to capture the knowledge from individual data and to ensure privacy protection in a user-centric scenario.

Personal data analytics and individual privacy protection are the key elements to leverage nowadays services to a new type of systems. The availability of personal analytics tools able to extract hidden knowledge from individual data while protecting privacy can help the society to move from organization-centric systems to user-centric systems, where the user is the owner of her personal data and is able to manage, understand, exploit, control and share her own data and the knowledge deliverable from them in a completely safe way.

The workshop is addressed to researchers interested on the advancement of personal data analytics, personal services development, privacy, data protection and privacy risk assessment, with a focus on issues related to personal analytics, personal data mining and privacy in the context where real individual data (spatio-temporal data, call details records, tweets, mobility data, social networking data, etc.) are used for developing a data-driven service, for realizing a social study aimed at understanding nowadays society, and for publication purposes.

The second edition of the International Workshop on Personal Analytics and Privacy (PAP 2018) has been held in Dublin, Ireland, in conjunction with the European Conference on Machine Learning and Principles and Practice of Knowledge Discovery in Databases (ECML-PKDD 2018). The format of the workshop included two keynote presentations and several technical presentations. The workshop has been attended by more than thirty people on average. The workshop received 6 submissions: after an accurate peer-review process, made in accordance with the reviewers in Program Committee, we selected 4 papers for presentation at the workshop, with an acceptance rate of 60%.

We would like to thank the PAP 2018 Program Committee, whose members made the workshop possible with their review process, and the ECML-PKDD conference for hosting the workshop.

Organization

PAP Chairs

Riccardo Guidotti	University of Pisa, Italy
Céline Robardet	Université de Lyon, France
Livio Bioglio	University of Turin, Italy
Ruggero G. Pensa	University of Turin, Italy

Program Committee

Luca Maria Aiello	Nokia Bell Labs, Cambridge, UK
Sonia Ben Mokhtar	LIRIS CNRS, Lyon, France
Francesco Buccafurri	University of Reggio Calabria, Italy
Paolo Cintia	University of Pisa, Italy
Jon Crowcroft	University of Cambridge, UK
Mathieu Cunche	University of Lyon/Inria, France
Daniele Dell'Aglio	University of Zurich, Switzerland
Boxiang Dong	Montclair State University, NJ, USA
Stefan Kramer	J. Gutenberg University Mainz, Germany
Bruno Lepri	FBK-Irst, Trento, Italy
Giuseppe Manco	ICAR-CNR, Rende, Italy
Michael Mathioudakis	University of Helsinki, Finland
Ioanna Miliou	University of Pisa, Italy
Richard Mortier	University of Cambridge, UK
Mirco Musolesi	University College London, UK
Francesca Pratesi	ISTI-CNR, Pisa, Italy
Rossano Schifanella	University of Turin, Italy
Andrea Tagarelli	DIMES, University of Calabria, Italy
Vicenc Torra	University of Skövde, Sweden
Roberto Trasarti	ISTI-CNR, Pisa, Italy

A Differential Privacy Workflow
for Inference of Parameters
in the Rasch Model

Teresa Anna Steiner, David Enslev Nyrnberg, and Lars Kai Hansen[✉]

Department of Applied Mathematics and Computer Science,
Technical University of Denmark B324, 2800 Kongens Lyngby, Denmark
{s170063,s123997}@student.dtu.dk,lkai@dtu.dk

Abstract. The Rasch model is used to estimate student performance and task difficulty in simple test scenarios. We design a workflow for enhancing student feedback by release of difficulty parameters in the Rasch model with privacy protection using differential privacy. We provide a first proof of differential privacy in Rasch models and derive the minimum noise level in objective perturbation to guarantee a given privacy budget. We test the workflow in simulations and in two real data sets.

Keywords: Rasch model · Differential privacy · Student feedback

1 Introduction

Protection of private information is a key democratic value and so-called 'privacy by design' is core to the new European General Data Protection Regulatory (GDPR) [6].

Privacy by design as a concept has existed for years now, but it is only just becoming part of a legal requirement with the GDPR. At its core, privacy by design calls for the inclusion of data protection from the onset of the designing of systems, rather than an addition. More specifically - 'The controller shall..implement appropriate technical and organisational measures..in an effective way.. in order to meet the requirements of this Regulation and protect the rights of data subjects'.

Differential privacy is one such tool allowing a 'controller' to train a machine learning model on inherently private data, but with mathematical bounds on the actual loss of privacy when results are released [4,8]. Differential privacy is based on randomized algorithms using noise to reduce the probability of breach of privacy. The key idea is to secure that the randomized output does not in a significant way depend on any of the possible data subjects' data.

This work was supported by the Danish Innovation Foundation through the Danish Center for Big Data Analytics and Innovation (DABAI).

C. Alzate et al. (Eds.): MIDAS 2018/PAP 2018, LNAI 11054, pp. 113–124, 2019.
https://doi.org/10.1007/978-3-030-13463-1_9

Educational technology is important to serve the increasing needs for life-long learning [10]. Learning processes and tests are typically highly personal, yet, significant gains are conceivable from integrating and sharing such information. Sharing could, e.g., be used to provide more detailed feedback on tests, hence, enhance the learning process. The basic question addressed in the present work is if differentially private machine learning methods can be used to provide more detailed feedback on students' tests, while still respecting the privacy of the individual students.

The concept is illustrated in Fig. 1. The use case concerns a class of students each answering a set of tasks. The teacher ('the controller') can by conventional means estimate each students performance and release this information in private to the given student. Here our aim is in addition to share a difficulty score for each task and investigate whether it is feasible to compute this score in a differentially private manner, hence, with mathematical bounds on the amount of individual information leaked by releasing the difficulty scores. Given the privatized difficulty scores, every student can then use their sensitive data to estimate their own ability scores and probabilities of passing a subject. The paper is organized as follows. We first present the differential privacy model in the educational technology context. Student performance and test scores are inferred using item response theory ('Rasch model'). Next, we investigate the loss of accuracy when privacy is enforced at various privacy budgets. Finally, we demonstrate viability in a real world data set. The proof of the differential privacy mechanism (so-called objective perturbation) is provided in an appendix. The original contributions can be summarized as follows: *(1) We define a workflow and model for privacy preserving machine learning of student performance and task difficulty. (2) We show by simulation that the student performance is well estimated for each student separately. (3) We give the first proof of differential privacy for the Rasch model based on so-called objective perturbation. (4) We derive the minimum noise level that allows us to release the task difficulty at a given privacy budget.*

All code can be found at the following github repository.

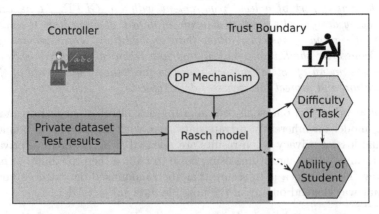

Fig. 1. Concept of the differentially private Rasch model and its use in enhanced feedback in teaching.

2   Preliminaries

The concept of differential privacy is based on a privacy parameter or 'budget' ϵ. The algorithm $\mathcal{A}$ is ϵ-differentially private if for all data sets D_1 and D_2 that differ by a single entry (data subject)

$$P[\mathcal{A}(D_1) = w] \leq P[\mathcal{A}(D_2) = w]\exp(\epsilon), \tag{1}$$

where P is the probability taken over the randomness used by the algorithm $\mathcal{A}$, and w is the output of the algorithm. The privacy budget quantifies how likely it is that a well-informed adversarial can determine whether a specific data subject participated or not. The randomness is added to the algorithm to hinder this identification. This randomness is achieved e.g. through the addition of noise. This noise is scaled as $\Delta f/\epsilon$ where Δf is the sensitivity of a function f, defined as

$$\Delta f = \max \|f(D_1) - f(D_2)\|_1, \tag{2}$$

where again D_1 and D_2 differ in a single entry [8].

The data sets we work with are arranged as a (number of students N) $\times$ (number of test items I) matrix X, and every entry stands for a right or wrong answer of a student to an item. In this work, we consider differential privacy in the sense that the output from our model should not depend much on whether a particular student is in the set or not. That is, for X and $\tilde{X}$ with $X_{n,i} = \tilde{X}_{n,i}$ for all $n = 1, \ldots, N-1$ and $i = 1, \ldots, I$, so two data sets that can differ in at most one row (corresponding to one student), we want to achieve

$$\frac{P(w|X)}{P(w|\tilde{X})} \leq e^\epsilon, \tag{3}$$

where w is the output of the algorithm.

The Rasch model [3] is a simple example of item response theory (IRT). IRT concerns performance testing quantifying the probability that students can answer a specific test task in terms of the difficulty of the task and their general ability. The model is similar to the logistic regression and used to estimate the probability of passing a task

$$P(X_{n,i} = 1|\beta_n, \delta_i) = \frac{\exp(\beta_n - \delta_i)}{1 + \exp(\beta_n - \delta_i)}, \tag{4}$$

where β_n models the ability of student n and δ_i is the difficulty of task i. X_{ni} is a dichotomous observation of a student's (n) correct or incorrect answer to a task (i), where 1 is a correct answer, and 0 is an incorrect answer. The model is generated by estimating δ_i and β_n from the results of a particular test. The parameters are estimated by maximizing the likelihood

$$\Lambda = \prod_n \prod_i \frac{e^{x_{ni}(\beta_n - \delta_i)}}{1 + e^{(\beta_n - \delta_i)}}. \tag{5}$$

In our work, we will introduce differentially private methods of estimating β and δ, but only the δ-values will be released to the public. Every student can then, based on the public δ and their personal results, estimate their own parameter β_n.

3   Methods

We will implement this workflow, i.e., release differentially private δ parameters and then re-estimate the parameters β_n on $X_{n,:}$, by first calculating both parameters with differentially private algorithms, assuring a private δ, then re-estimating each β_n given that δ. The re-estimation was also proposed in Choppin [3]. We will investigate the impact of the re-estimation compared to the global parameter estimation in Sect. 4.

We consider two different methods for constructing a differentially private Rasch Model. The first one is the objective perturbation, first introduced by Chaudhuri and Monteleoni for logistic regression [1], and analyzed in more detail by Chaudhuri et al. [2]. For the differentially private Rasch model, we use a slightly modified version of the objective perturbation, and prove that it is ϵ-differentially private as defined in Eq. (3).

We will also consider a simple reference method based on perturbing sufficient statistics as discussed in [7]. By adding enough noise to the sufficient statistics, we may release them as differentially private. Then any algorithm based on these statistics will be differentially private. The latter observation follows from the post-processing theorem [5]. For the Rasch model, the sufficient statistics are $r_n = \sum_i^I X_{n,i}$ and $s_i = \sum_n^N X_{n,i}$, since those are all that is needed for minimizing the regularized objective function. We add noise to the vectors r and s, scaled with their sensibilities: if student n in the data set is changed, r_n will change by at most I, and so the L_1 norm of r will change by at most I. Similarly, s_i can change by at most 1 for every i, so again, the L_1 norm of s changes by at most I. So the noise we add to both vectors is scaled with I/ϵ as in [7]. Since making the sufficient statistics differentially private is more general than objective perturbation, and does not use the specific structure of the learning algorithm, we expect it to be weaker (adding more noise).

As we notice below, the objective perturbation approach effectively perturbs the sufficient statistics with noise scaling as $\sqrt{I}/\epsilon$, while direct perturbation of the sufficient statistics requires noise scaling as I/ϵ [7]. In the following we will show the costs of the less favorable scaling.

An important aspect of learning the Rasch model parameters is to quantify the available prior information. In the educational technology context we could imagine substantial prior information to be present from earlier exams etc. Here we for simplicity assume that the test difficulties and student abilities both follow normal distributions, hence, we add a regularization term $\frac{\lambda}{2} w^T w$, where w is the entire parameter vector $w = [\beta \ \delta]$, to the Log likelihood function in equation (5).

A discussion on how to estimate parameter λ while preserving differential privacy can be found below.

Algorithm 1 describes the details of the modified objective perturbation algorithm for the differentially private Rasch model.

Algorithm 1:

- Draw vector b with dimension I from density function $h(b) \propto \exp\left(-\frac{\epsilon}{\sqrt{I}}\right)$. To do that, draw direction uniformly at random from the I dimensional unit sphere, and draw the norm from $\Gamma(I, \frac{\sqrt{I}}{\epsilon})$.
- Minimize

$$
F(\beta, \delta) = \sum_n^N \sum_i^I \log(1 + \exp(\beta_n - \delta_i)) + \sum_i^I \delta_i \sum_n^N X_{n,i}
$$
$$
- \sum_n^N \beta_n \sum_i^I X_{n,i} + \frac{\lambda}{2}(\beta^T \beta + \delta^T \delta) + \sum_i^I b_i \delta_i \tag{6}
$$

with respect to β, δ for $\lambda > 0$.

Theorem 1. *Algorithm 1 is ϵ-differentially private.*

The proof follows the strategy developed in Chaudhuri et al. [1,2], details are found in the appendix.

Naive approaches of estimating the regularization parameters by f.e. cross validation can lead to a loss of privacy, since information is leaked by every evaluation of the model, and this information accumulates. Chaudhuri et al. [2] propose two different methods on how to handle this, which we can also apply. The first one is to use a small publicly available data set that follows roughly the same distribution for the estimation of λ. The second one is an algorithm which splits the data set into $(m + 1)$ subsets, calculates the model for m different guesses of λ on respective subsets, and evaluates the model on the last one. Then, based on the number z_i of errors made by the ith estimate, λ_i is chosen with probability

$$
\frac{e^{-\epsilon z_i/2}}{\sum_{j=1}^m e^{-\epsilon z_j/2}}. \tag{7}
$$

For our model, we split the data into subsets of students. For the m different estimates of δ, we first compute the β values for the last subset and the corresponding probabilities. We then compare the rounded probabilities to the actual 0 or 1 entries in the $(m + 1)^{\text{st}}$ subset in the data.

4   Experiments

Experiments were run for simulated data, real data from Vahdat et al. [9], as well as a data set from a course at the Technical University of Denmark.

The first experiment is run on simulated data and compares the results of calculating both β and δ globally to re-estimating β.

The next experiments compare the performance of the two methods for introducing privacy, the objective perturbation and sufficient statistics. We use correlation coefficients between the estimated probabilities and the true probabilities (i.e., the ones used to simulate the data) resp. the non-private estimates with 95% confidence intervals. Further, we show test misclassification rates (how well do we predict if a student passes a test) on a new data set drawn from the same distribution as the training data. Experiments were run for the two real data sets. The first one had to be modified to fit the Rasch model, so the answers were rounded to 1 or 0 depending on whether half of the points for a question were scored. The DTU data set are the results of a multiple choice test, so the original data can be used. For the real data sets, we use bootstrapping with 1000 samples to calculate confidence intervals.

We used MATLAB's *fminunc* function for minimizing the objective function in all experiments - with the following settings:

```
options = optimoptions('fminunc','Algorithm','quasi-newton',
'off','SpecifyObjectiveGradient',true,'MaxIter',10^5,
'MaxFunEvals',10^5,'TolX',10^-5);
```

The experiments on simulated data were run with 50 repetitions, used a regularization parameter of 0.01, and privacy budgets of 1, 5 and 10. The amount of students (N) vary from 40 to 200 in steps of 40 and amount of questions (I) is fixed to 20. The parameters β_n and δ_i were drawn from normal distributions with mean 0, and standard deviation 1 and 2, respectively. The Rasch probabilities were then calculated with the given β_n and δ_i and used to simulate a data set by drawing from a Bernoulli distribution with the given probabilities.

The M. Vahdat et al. data set has 62 students and 16 questions. The DTU data set has 212 students and 27 questions.

In pilot experiments we found that fine-tuning of the regularization parameter λ was not essential. So for simplicity reasons we use a common regularization parameter in all experiments.

4.1   Results

To test the impact of introducing differential privacy to the Rasch model, we ran several experiments. In the first one we test the retraining of β in the non-private setting, to show the students can calculate their own abilities from a given, private δ. The second experiment show the performance of the two methods on simulated data, while the third and fourth show the performance on the M. Vahdat real data and DTU data, respectively.

Experiment 1: Global Vs. Re-estimated Rasch Parameters. In Fig. 2 we compare the Rasch model with global parameter estimation with the results obtained by re-estimating the student abilities. We plot correlation coefficients with 95% confidence intervals between the probabilities, also with respect to the true model parameters, as well as misclassification rates on a new data set drawn from the same distribution as the train data.

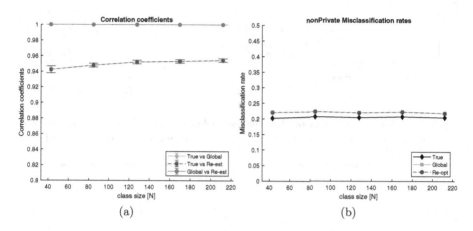

Fig. 2. The plots show: (a) Correlation coefficient for non-private global and re-estimation method compared to the true Rasch model. (b) Misclassification for non-private global and retrain estimation method compared to the true Rasch model.

From the correlation coefficients between the estimated probabilities and the ground truth probabilities used to simulate the data, we can see that there is practically no difference in accuracy between re-estimation and global estimation. From now on, we will only consider the re-estimation method, since this is the one defined by our workflow.

Experiment 2: Differential Privacy on Simulated Data. In Fig. 3 we show a comparison of the objective perturbation and sufficient statistics method with three different values of epsilon: 1, 5 and 10. We compare their performance by calculating the correlation coefficients between the private estimation to the non-private resp. true estimation. Again, we show 95% confidence intervals.

In Fig. 4, we show the missclassification rates on a simulated data set of the same distribution as the training set.

We make several observations. First, the objective perturbation performs better in general. This is due to the smaller amount of noise added to the objective perturbation, as mentioned in Sect. 3. Next, we see that for lower epsilon values, i.e. higher privacy, the model generally performs worse, but converges with larger class size. This is what we would expect and consistent with what is broadly observed in applications of differential privacy.

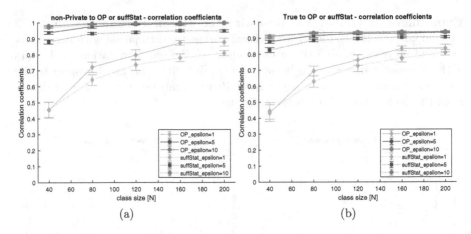

Fig. 3. Correlation coefficient of objective perturbation and sufficient statistics methods with $\epsilon = (1, 5$ and $10)$ to: (a) the non-Private estimates. (b) the true model.

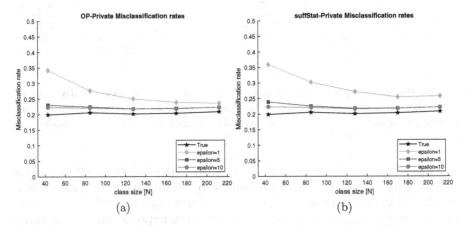

Fig. 4. Misclassification rates of objective perturbation and sufficient statistics method with $\epsilon = (1, 5$ and $10)$: (a) objective perturbation estimates. (b) sufficient statistics estimates.

Experiment 3 and 4: Differential Privacy on Real Data. Experiment 2 illustrated the impact of privacy so for experiments 3 and 4, the privacy budget was fixed to $\epsilon = 5$ with changing data sizes. In experiment 3 we use the data set from Vahdat et al. [9]. Experiment 4 is run on the DTU data set.

Figure 5 shows the objective perturbation and sufficient statistics performance on real data with $\epsilon = 5$. The misclassification rate here is calculated on the original data set (so corresponds to a train, not a test error).

In experiment 3, Fig. 5 (a) and (b), we show that the impact of introducing privacy on real data sets is limited, even in relatively small data sets. For both

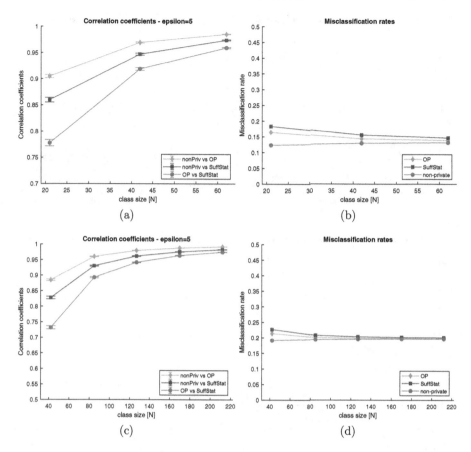

Fig. 5. Objective perturbation and sufficient statistics methods on real data: (a) Correlation coefficients on Vahdat et al.'s data [9]. (b) misclassification on Vahdat et al.'s data [9]. (c) Correlation coefficients on DTU data. (d) misclassification on DTU data.

the Vahdat et al. data and the DTU data we find useful correlation between the probabilities of passing the test as inferred in non-private and private models ($\epsilon = 5$).

In experiment 4, Fig. 5 (c) and (d), our results are comparable to those on the simulated data. In general, we see that the objective perturbation method performs better than the sufficient statistics method.

In Fig. 6, we illustrate the impact of data set size by showing comparisons of the probability estimates of the non-private model to the two private methods on the data set from Vahdat et al. [9], again with fixed $\epsilon = 5$.

We see how the estimates are very noisy for small data set sizes, but correlate strongly with the non private for a data set size of 62. Again, the objective perturbation yields more accurate results.

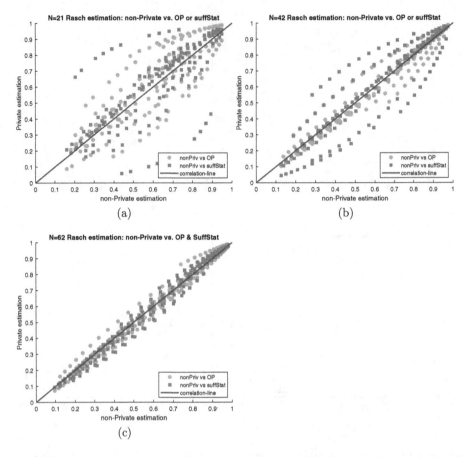

Fig. 6. Rasch estimates of objective perturbation and sufficient statistics methods on Vahdat et al.'s data [9]: (a) nStudent = 21. (b) nStudent = 42. (c) nStudent = 62

5   Conclusion

We have demonstrated viability of the proposed workflow for more detailed, yet differentially private, feedback for students. We proved analytically that objective perturbation for this model satisfies differential privacy and give the minimum noise level necessary. Our experiments based on simulated data suggest that the workflow provides estimates of similar quality as the non-private for medium sized classes and industry standard privacy budgets[1]. These findings were confirmed in two real data sets. As expected, the objective perturbation mechanism performs better than the sufficient statistic method as less noise is added.

---

[1] See e.g., https://www.wired.com/story/apple-differential-privacy-shortcomings/.

Acknowledgements. We would like to thank Martin Søren Engmann Djurhuus, who worked with us on the project in its early stages during the course "Advanced Machine Learning" at DTU. Further, we would like to thank Morten Mørup for access to the DTU data.

Appendix

Proof (of Theorem 1). Since the objective function is differentiable everywhere and a minimizing pair (β^*, δ^*) satisfies $\nabla F(\beta^*, \delta^*) = 0$, for every output (β^*, δ^*) there exists exactly one b which maps the input to the output. On the other hand, since the objective function (6) for $\lambda > 0$ is strongly convex (which can be seen by computing the Hessian matrix H and realizing that $H - \lambda I$ is positive semidefinite), for any fixed b and X, there is exactly one pair β^*, δ^* which minimizes the function. As such there is a bijection between (β^*, δ^*) and b.

Now consider two data sets X and $\tilde{X}$ that differ in exactly one student (w.l.o.g., the last one). For β^*, δ^* minimizing (6) for both X and $\tilde{X}$ denote the corresponding noise vectors b and $\tilde{b}$. By the transformation property of probability density functions, we get

$$\frac{P(\delta^*, \beta^*|X)}{P(\delta^*, \beta^*|\tilde{X})} = \frac{h(b)\left|\det\left(J_{(\beta^*, \delta^*, \tilde{X})}(\tilde{b})\right)\right|}{h(\tilde{b})\left|\det\left(J_{(\beta^*, \delta^*, X)}(b)\right)\right|}, \tag{8}$$

where $J_{(\beta^*, \delta^*, X)}(b)$ denotes the Jacobian matrix of b as a function of (β^*, δ^*), given input set X.

We get b_i as a function of (β^*, δ^*) by setting the gradient of the objective to zero:

$$b_i = \sum_{n=1}^{N} \frac{e^{\beta_n^*}}{e^{\beta_n^*} + e^{\delta_i^*}} - \sum_{n=1}^{N} X_{ni} - \lambda \delta_i^*. \tag{9}$$

Since the sum over X will disappear in any derivative for δ_i^* and β_n^*, the Jacobian matrices in (8) are identical and the determinants cancel.

Furthermore, since $X_{ni} = \tilde{X}_{ni}$ for all $i = 1, \ldots, I$ and all $n = 1, \ldots, (N-1)$, equation (9) also gives

$$X_N + b = \tilde{X}_N + \tilde{b}.$$

By the reverse triangle inequality we get

$$\left|\|b\| - \|\tilde{b}\|\right| \le \|b - \tilde{b}\| = \|X_N - \tilde{X}_N\| \le \sqrt{I}$$

and thus

$$\frac{P(\delta^*, \beta^*|X)}{P(\delta^*, \beta^*|\tilde{X})} = \frac{h(b)}{h(\tilde{b})} = e^{-\frac{\epsilon}{\sqrt{I}}(\|b\| - \|\tilde{b}\|)} \le e^{\epsilon} \tag{10}$$

$$\square$$

References

1. Chaudhuri, K., Monteleoni, C.: Privacy-preserving logistic regression. In: Proceedings of the 21st International Conference on Neural Information Processing Systems, NIPS 2008, pp. 289–296. Curran Associates Inc., USA (2008). http://dl.acm.org/citation.cfm?id=2981780.2981817
2. Chaudhuri, K., Monteleoni, C., Sarwate, A.D.: Differentially private empirical risk minimization. J. Mach. Learn. Res. **12**(Mar), 1069–1109 (2011)
3. Choppin, B.: A fully conditional estimation procedure for Rasch model parameters (CSE report 196): University of California. Center for the Study of Evaluation (1983)
4. Dwork, C.: Differential privacy: a survey of results. In: Agrawal, M., Du, D., Duan, Z., Li, A. (eds.) TAMC 2008. LNCS, vol. 4978, pp. 1–19. Springer, Heidelberg (2008). https://doi.org/10.1007/978-3-540-79228-4_1
5. Dwork, C., Roth, A., et al.: The algorithmic foundations of differential privacy. Found. Trends® Theor. Comput. Sci. **9**(3–4), 211–407 (2014)
6. EU GDPR Portal: GDPR key changes - an overview of the main changes under GDPR and how they differ from the previous directive. https://www.eugdpr.org/key-changes.html (2018). Accessed 19 May 2018
7. Foulds, J., Geumlek, J., Welling, M., Chaudhuri, K.: On the theory and practice of privacy-preserving bayesian data analysis. In: Proceedings of the Thirty-Second Conference on Uncertainty in Artificial Intelligence, pp. 192–201. AUAI Press (2016)
8. Ji, Z., Lipton, Z.C., Elkan, C.: Differential privacy and machine learning: a survey and review. arXiv preprint arXiv:1412.7584 (2014)
9. Vahdat, M., Oneto, L., Anguita, D., Funk, M., Rauterberg, M.: A learning analytics approach to correlate the academic achievements of students with interaction data from an educational simulator. In: Conole, G., Klobučar, T., Rensing, C., Konert, J., Lavoué, É. (eds.) EC-TEL 2015. LNCS, vol. 9307, pp. 352–366. Springer, Cham (2015). https://doi.org/10.1007/978-3-319-24258-3_26. https://archive.ics.uci.edu/ml/machine-learning-databases/00346/
10. Uden, L., Liberona, D., Welzer, T.: Learning technology for education in cloud. In: Third International Workshop. LTEC 2014. Springer (2014)

Privacy Preserving Client/Vertical-Servers Classification

Derian Boer, Zahra Ahmadi[✉], and Stefan Kramer

Institut für Informatik, Johannes Gutenberg-Universität,
Staudingerweg 9, 55128 Mainz, Germany
dboer@students.uni-mainz.de, {zaahmadi,kramer}@informatik.uni-mainz.de

Abstract. We present a novel client/vertical-servers architecture for hybrid multi-party classification problem. The model consists of clients whose attributes are distributed on multiple servers and remain secret during training and testing. Our solution builds privacy-preserving random forests and completes them with a special private set intersection protocol that provides a central commodity server with anonymous conditional statistics. Subsequently, the private set intersection protocol can be used to privately classify the queries of new clients using the commodity server's statistics. The proviso is that the commodity server must not collude with other parties. In cases where this restriction is acceptable, it allows an effective method without computationally expensive public key operations, while it is still secure and avoids precision losses. We report the runtime results on some real-world datasets, and discuss different security aspects and finally give an outlook on further improvements.

Keywords: Vertically partitioned data · Private evaluation ·
Secure multi-party computation · Privacy preserving data mining ·
Random forest

1 Introduction

In the era of big data, the costs of storing and processing data is decreasing and as a result, the amount of collected data for analyzing purposes is also increasing. In this context, the goal is to make use of the potential knowledge, which additionally measured or mined data can provide. At the same time, the challenges of preserving privacy of personal and other sensitive data grow. This challenge becomes more important especially when partners collaborate with the intention to benefit from the union of their data. These partners can also be competitors, for example, and more or less trusted. Legal privacy concerns have to be considered as constraints in these cases. Aspects of privacy-preserving data include randomization, k-anonymity and l-diversity, downgrading application effectiveness, and secure multi-party computing (SMC) [1]. While randomization, k-anonymity and effectiveness downgrading require a trade-off between effectiveness (quality of the output) and privacy, SMC techniques do not effect

© Springer Nature Switzerland AG 2019
C. Alzate et al. (Eds.): MIDAS 2018/PAP 2018, LNAI 11054, pp. 125–140, 2019.
https://doi.org/10.1007/978-3-030-13463-1_10

the effectiveness. In SMC, the data sets remain completely private and hence need not be modified. Instead, special cryptographic communication protocols allow two or more parties to obtain aggregated results from their combined data, but each of them does not learn any information more than what can be derived from their common output. Thereby, the SMC algorithms lead to the same results as non-private algorithms do. This research area is also known as distributed privacy preservation, because the integrated data are partitioned on multiple parties who protect their shares. A special and upcoming case of secure multi-party computing is private evaluation, where a server has a sensitive model and a client has sensitive attribute vectors as input. The goal is that the client obtains a classification of her attribute vector with the use of the server's model, while the client does not see the model in plain text and the server is not able to get any parts of the client's input and output.

In this work, we consider a special case of the aforementioned private evaluation scenario for decision tree-based classification and set intersection in the secure multi-party computation scenario. We assume that the data of some clients is vertically partitioned and distributed across multiple servers. It is sufficient if at least one server knows the class values of the training instances. Each party can only see her attributes of the common decision trees. The leaf node statistics are stored by a trusted third party, which does not know any instances or tree attributes. To new test instances are classified with the following steps:

- The attribute vector of the client is vertically partitioned on the servers.
- The servers run a novel private set intersection protocol so that the client gets a shared sum of her leaf node identifier from each party. No server learns other attributes or information about the output at this step.
- Then a client can anonymously ask the commodity server about the class value statistics of her leaf node.

Our proposed architecture benefits from the following features:

1. A fast computation run time by removing computationally expensive primitives that are used in public key encryption methods such as Vaidya and Clifton's method [26],
2. Linear scaling with respect to the number of parties,
3. Accurate results in contrast to randomization based techniques [12,14],
4. Similar to private evaluation methods, no other party is able to learn any node statistics and cannot detect the similarities among different records, however, in contrast to them, our framework can handle distributed decision trees.

Moreover, there are several applications for this framework in spam filtering, crime reduction or credit assessment. Police forces, tax authority and financial institutions might be willing to cooperate in terms of fraud prevention, but only want to share uncoded personal data in case of reasonable suspicion. Another typical application is clinical diagnosis. In a real-world setting, the sensitive data of several institutions might be necessary to come up with a good diagnosis for a client or a responsible expert, while none of them want to disclose their information to the others.

The rest of the paper establishes the proposed secure architecture in details. First, we give an overview on the background in the next section. Section 3 presents the problem setting and the proposed client/vertical server random forest model. We elaborate experimental results in Sect. 4. Finally, Sect. 5 concludes the paper.

2   Background

In the multi-party computing scenario, data can be partitioned among parties vertically, where the parties have different attributes of the same data objects, or horizontally, where the parties have different data objects of a compatible structure. There have been extensive studies on both partitioning approaches for privacy preserving decision tree based methods. In the rest of this section, we first provide an overview of decision tree induction and random forests, then, we briefly review the private decision tree learning literature for both vertically distributed data and private evaluation. As the focus of our work lies only on vertically partitioned data and private evaluation, we discard reviewing the horizontal private decision tree approaches [11,18,20].

ID3 and Random Forest: Decision trees are commonly used not only for solving classification and regression problems, but also for clustering with cluster descriptions [2]. A widely-used, intuitive decision tree algorithm is the ID3 algorithm [21] and its extension C4.5. Both use the Shannon entropy and information gain to create tree branches efficiently. The entropy can be replaced by other impurity measures with different sensitivities to costs [9]. Decision trees have shown promising results on many problems, nevertheless, their performance can be improved by a majority vote that combines the outcomes from many largely independent decision trees. A single tree is inclined to overfitting which causes a high variance. The random forest algorithm [5] tackles this instability by building several trees based on two randomization concepts: First, the training sets are varied by bootstrap aggregating, an equally-distributed random selection of records with replacement. Second, at each splitting step, it selects only s random attributes. This has the additional advantage that the trees can be constructed with very few or even without any data queries in the first step, if s is set to one. As a successful and well-interpretable learning model, in this paper, we focus on secure multi-party computation models for decision trees.

Secure Multi-party ID3 on Vertically Partitioned Data: The first work on vertically partitioned data for two parties was proposed by Du and Zahn [10]. In order to count the records that support a particular attribute or class value, they suggest that every party fills a binary vector with a one, if a record conforms with the currently examined attributes, and a zero, if a record does not. A secure shared two-party scalar product protocol and a secure shared logarithm protocol on these vectors allow to calculate the conditional entropy without revealing the involved records. Similar to our architecture, these protocols are lightweight solutions, and require a commodity server that should not collude with any of the parties. However, this approach is hardly extensible to an n-party solution. The other drawback is the possibility of revealing sensitive data by making inferences from the public decision tree.

To solve the inference problem, some approaches use trees whose nodes are mapped to the attributes that are only visible for the corresponding party [23,26]. Vaidya and Clifton propose a private set intersection protocol (PSIP) [26], which can be applied in the same way as the scalar product protocol [10]. Their PSIP is based on public homomorphic encryption [8], and it requires a high number of key bits which increases the runtime significantly. Recently, some private set intersection protocols that use symmetric key encryption have been developed [16,17], however, the communication costs still remain relatively high. In another approach [23], each party finds the attributes with the highest information gain independently from the other parties at each branch step. The party with the attribute of the highest gain executes the split and broadcasts the separation of the records, but not the identity of the split attribute. This approach is only feasible if each party holds the class attribute. The other downside of this approach is that the similarities among different records can still be leaked. Both of these two approaches can support two or several parties.

Decision Trees on Randomized Data: Randomization approaches usually lead to faster results, but imply a trade-off between the individual's privacy and the quality of the results. The first randomization based multi-party tree induction used a multi-group randomized response (MRR) scheme [27] that works as follows: The attributes are partitioned into groups. In a first step, coin-flipping is conducted for each group and a user either tells the truth about all attributes in the same group or tells a lie about all of them. The trade-off between privacy and performance is regulated by a fixed probability of lie. One party works as a data collector of all randomized data sets and executes the ID3 algorithm on the collection. The cost of tree building and the accuracy loss can be reduced by employing a hybrid of MRR and SMC [25]. The ω attributes that have the highest information gains on the combined randomized data are selected and evaluated on a private dot product or intersection protocol.

Recently, many ϵ-differential solutions for ID3 [12,15] and tree ensembles [3, 14,19,22] were proposed. A solution is ϵ-differential private if the outcome of a calculation is insensitive to any particular record in the data set:

Definition 1. *A randomized computation M provides ϵ-differential privacy, if for any datasets A and B with symmetric difference $A \triangle B = 1$, and any set of possible outcomes $S \subseteq Range(M)$, $Pr[M(A) \in S] \leq Pr[M(B) \in S] \times e^{\epsilon}$.*

The randomization is obtained by the addition of noise, and the ϵ-parameter can be seen as a "privacy budget". As it fulfills the composability property, the parameters of consecutive queries can be accumulated. The major drawback, especially of the ID3 solutions, is the high variance in the accuracies, and that only the individual's privacy is preserved, but not the private attribute distributions. Therefore, random decision trees have been shown to be more efficient and provide better security than ID3 induced trees in the context of differential privacy [22].

Private Tree Evaluation with a Client-Server Architecture: The field of privacy-preserving decision tree evaluation is a different, yet somewhat related to what we already discussed. Here, a server has a sensitive model and a client has sensitive attribute vectors as input. The goal is to classify the client's data while the sensitive inputs (model and query) remain hidden from the counterparty. Many approaches use homomorphic encryption [4,6,13,24]. Wu *et al.* [4] and Tai *et al.* [24] reduce the protocol complexity to be linear with respect to the number of decision nodes by representing the decision trees, which are high-degree polynomials, as linear functions.

3 Client/Vertical-Servers Random Forest

In this section, we explain our proposed client-server architecture in detail. First, we establish the problem definition. Then, we explain the private set intersection protocol, and how to build and apply the decision tree in a client-server architecture. Finally, we analyze the security of our proposed architecture.

3.1 Problem Setting

Problem Description: In this work, we consider the classification problem in the hybrid context of vertically partitioned data and the client-server architecture. Our proposed architecture is composed of two modes: a training phase in which the decision tree is built, and an evaluation mode where a test instance of a client is classified by the existing private model. In the training phase, we have m input vectors (training instances), whose attributes are distributed on n vertical servers p_i. Let $X_i[j]$ denote the attribute values of the record j that is known by p_i. The target value is held by the class server p_c, which can be one of the vertical servers, individual clients, or any other party. In the evaluation phase, one client c's query is classified by a trained model $f(x)$.

Constraints and Assumptions: The classification is a private service, such that no vertical server p_i is able to reveal any information about the attributes ($X_i[j]$) and the target value of c or similarities to other clients or training records. The client should not learn anything about the underlying model or any $X_i[j]$ than what can be deduced from $f(c)$. We allow the use of a semi-honest commodity server cs, which must not collude with any p_i and receives nothing but anonymous data.

Algorithm 1. Client/Vertical-Servers Set Intersection (CVSSI)

> **input** : A vector $D_i \in \{0,1\}^s$ for each server p_i, $i \in \{1, ..., n\}$,
> a collusion threshold parameter T
>
> **output:** A vector $Y \in \{0,1\}^s$, where $Y[j] = \prod\limits_{i=1}^{n} D_i[j]$

```
1  for j ← 1 to s do
2  |   foreach server pi do
3  |   |   ri⁰[j] ← new random number
4  |   |   for t ← 1 to T do
5  |   |   |   rit[j] ← new random number
6  |   |   |   send rit[j] to server pi+t mod n
```

$$7 \quad R_i[j] \leftarrow \sum_{t=0}^{T} r_i^t[j]$$

```
8  |   |   send Ri[j] to client
9  |   |   for t ← T to 1 do
10 |   |   |   receive r^t_{i-t mod n}[j] from server pi-t mod n
```

$$11 \quad S_i[j] \leftarrow \begin{cases} \sum\limits_{t=0}^{T} r_{i-t \ mod \ n}^t[j] & \text{if } D_i[j] = 0 \\ \text{new random number} & \text{else} \end{cases}$$

```
12 |   |   send Si[j] to client
13 |   for client do
```

$$14 \quad Y[j] \leftarrow \begin{cases} 1 & \text{if} \sum\limits_{i=1}^{n} (R_i[j] - S_i[j]) = 0 \\ 0 & \text{else} \end{cases}$$

3.2 Client/Vertical-Servers Set Intersection Protocol

We provide a special variant of a private set intersection as a major building block of both our model training and classification protocol. Assume that each vertical server p_i contains a binary vector $D_i \in \{0,1\}^s$, where $D_i[k] = 1$ if p_i supports the element k, and $D_i[k] = 0$, otherwise. The output is a binary vector $Y \in \{0,1\}^s$ with

$$Y[j] = \prod_{i=1}^{n} D_i[j]. \tag{1}$$

No vertical server p_i is allowed to reveal any value of D_j, $j \neq i$ or $f(c)$. Let T be a collusion threshold parameter. If less than T servers collude with each other, they can only exchange their inputs, but cannot induce input values of any other server or $f(c)$. A client cl receives the output, but should not learn anything else. Algorithm 1 solves this problem and calculates each $Y[j] \in Y$ independently from the others via a zero-sharing method. In zero-sharing, the servers distribute random numbers that sum up to zero. Then, every p_i sends its shares to cl if $D_i[j] = 1$.

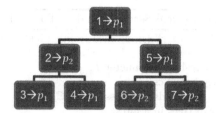

Fig. 1. Example of a random tree skeleton.

Table 1. Example of privately labeled tree.

Vertical server	Node	Attribute
p_1	1	Outlook
p_1	3, 4, 5	Humidity
p_2	2, 6, 7	Overcast

Algorithm 1 explains the steps of checking whether an element d is supported by all vertical servers in detail. All arithmetics are modulo integer operations in a sufficiently large field with the bit length b and all random numbers are uniformly distributed within this field. First, every p_i generates $T + 1$ random numbers, $\{r_i^0, \ldots, r_i^T\}$, where $T \in [1, n-1]$ (lines 3–5), sends their sum to cl (lines 7–8) and scatters $\{r_i^1, \ldots, r_i^T\}$ to T other servers (line 6 and 10). Then, every p_i sends the sum of $r_i^0[j]$ and all received values to cl, if it holds d; otherwise, it sends a random number (lines 11–12). The client cannot distinguish between this random number or the sum, which is composed of other random numbers. The client adds up all values it received in the first round and subtracts all the values that were received in the second round (line 14). If the result is zero, d is not held by all servers with certainty, but in case the results do not sum up to zero, all the servers hold d with a probability of $1 - 1/2^b$. The probability of a false positive is therefore negligible. For our classification purposes, the size of an intersection should be always one (see Sects. 3.3 and 3.4). Hence, the occurrence of a false positive can be detected easily and the procedure restarts for the candidate elements with new random seeds.

3.3   Building a Privacy-Preserved Random Forest

Algorithm 2 presents the training steps of a private random forest. The model is distributed over all vertical servers and a commodity server. Each one receives the same tree skeletons similar to the model already proposed by other authors [23, 26] (lines 1–3). It maps an identifier and a party p_i to each branch in a random assignment process. Every server maps selected attributes to each node randomly (lines 4–6). Figure 1 illustrates a sample tree skeleton and Table 1 indicates its privately labeled tree.

As the tree is built randomly, no data records have been used yet. In the following steps, only the commodity server cs learns the class value distributions

---

Algorithm 2. Client/Vertical-Servers Random Forest Training

---

input : An attribute vector $X_i[j]$ for each training instance $j \in \{1, ..., m\}$ of
each vertical server $\mathsf{p}_i, i \in \{1, ..., n\}$,
a collusion threshold parameter T,
a commodity server cs,
a server p_c, that holds the class values $C[j] \in \mathrm{C}$,
the number of trees in a forest o

output: A mapping $M : \mathbb{N} \mapsto \mathrm{C}$, that maps the index of each leaf node to the
most associated class value

1 **for** one arbitrary server **do**
2   **for** $k \leftarrow 1$ **to** o **do** // for each tree of random forest do
3        $\text{tree}^k \leftarrow$ new RandomTreeSkeleton()

4 **for** $k \leftarrow 1$ **to** o **do** // for each tree of random forest do
5   **foreach** vertical server p_i **do**
6        $\text{tree}_i^k \leftarrow \mathsf{p}_i.\texttt{labelPrivately}(\text{tree}^k)$

7 **for** $j \leftarrow 1$ **to** m **do** // for each record do
8   **for** $k \leftarrow 1$ **to** o **do** // for each tree of random forest do
9   **foreach** vertical server p_i **do**
10          $D_i \leftarrow \mathsf{p}_i.\texttt{getAllCandidateLeafs}(X_i[j], \text{tree}_i^k)$
11   leafID $\leftarrow$ CVSSI($\{D_i\}$, T)
12   cs.store(leafID, $\mathsf{p}_c.\texttt{getClassValue}(C[j])$)

---

of the leaves. It receives an assignment of a leaf node to each instance for each decision tree of the random forest. As long as cs does not collude with any vertical server, it cannot associate any attribute or class label with the identity of any instance. Algorithm 1 generates the leaf node that is provided to cs and the commodity server is treated as a client in this context. First, a preprocessing step is executed and each p_i assigns $D_i[l]$ with a one if the instance j reaches the leaf l (based on the attributes in $X_i[j]$), or a zero if not (line 10). Let $\{D_i\}$ denote the collection of private vectors D_i of all p_i. We get the output of Algorithm 1 for the record j, which is the leaf ID, $leafID$ (line 11). Then the commodity server updates the class distribution statistics of $leafID$ with $f(j)$ which is received from the class server p_c (line 12).

3.4   Client/Vertical-Servers Random Forest Classification

To classify a test instance c for a client whose attributes are stored at the vertical servers, we apply Algorithm 3. For that, all vertical servers p_i need to initialize a vector D_i with one at $D_i[l]$ if c reaches the leaf l or with zero, otherwise (line 3). Client c conforms to cl in Algorithm 1, so it receives the $leafID = c.leafIDs_{tree}$ corresponding to its attributes for each decision tree (line 4). Subsequently, it sends a request with all the leaf IDs to the commodity server to receive the most likely class label.

Algorithm 3. Client/Vertical-Servers Random Forest Classification

---

input : an attribute vector $X_i[c]$ of each vertical server $\mathsf{p}_i, i \in \{1, ..., n\}$,
a collusion threshold parameter T,
a commodity server cs with a mapping $M : \mathbb{N} \mapsto \mathrm{C}$, that maps the
index of each leaf node to the most associated class value,
a client c,
an ensemble of $o \times n$ decision trees $\{\mathsf{tree}_i^k\}$
output: Classification of c

---

1 **for** $k \leftarrow 1$ **to** o **do** // for each tree of random forest do
2   **foreach** vertical server p_i **do**
3        $D_i \leftarrow \mathsf{p}_i.\texttt{getAllCandidateLeafs}(X_i[c], \mathsf{tree}_i^k)$
4     $\mathsf{c.leafIDs}_{\text{tree}} \leftarrow \texttt{CVSSI}(\{D_i\}, T)$
5 $\mathsf{c.classValue} \leftarrow \texttt{cs.classify}(\mathsf{c.leafIDs})$

Note About the Client-Commodity Server Communication: The communication between c and cs is straightforward. Note that cs can read the client's request in clear text, but the client can communicate with the commodity server anonymously, so that cs cannot link the request with any other sensitive data or the identity of c. This communication can be done via a string-select oblivious transfer protocol, so that the commodity server does not learn the input of the client ($leafIDs$) and the output of the protocol. Kolesnikov *et al.* [17] provide an efficient 1-of-n oblivious transfer protocol, which can be applied here. It requires roughly four times the costs of a 1-out-of-2 oblivious transfer in an amortized setting and, therefore, is highly scalable. Moreover, c should get a shared one-time-password by one or more parties to prevent it from sending multiple malicious requests to cs, and not to be able to deduce sensitive information about the model and the underlying data. If these passwords are generated by the vertical servers, cs cannot associate them with individual clients even if the communication is not oblivious.

3.5   Security Analysis

In this section, we analyze the robustness of Algorithms 1–3 to information leakage. In Algorithm 2 (line 11) and 3 (line 4), the interactions among servers are limited to the interactions in Algorithm 1, hence, the security aspects of Algorithm 1 are directly transferable to them. Assuming that cs does not collude with any other server, the leaf IDs, class labels and input of vertical servers are secure against semi-honest and malicious attacks. The security level of the communication between the client and the commodity server (line 5 of Algorithm 3) is adaptable as discussed in Sect. 3.4. Here, we will discuss different security aspects of Algorithm 1:

– **Disclosure of the output:** The goal of Algorithm 1 is that the vertical servers input their private sets and the commodity server receives the intersection as the output. Assume that one or more parties try to reveal information about the output. The only messages they receive are random based zero-shares of other servers, which are independent from both their own input and any input of other parties. Consequently, even malicious parties have no opportunity to disclose anything about the output.

– **Association of input and output:** In case of a collusion between the commodity server and a vertical server, the collaborators can associate the identities of all the training records to their corresponding sensitive class values and leaf nodes, and therefore, similarities between the records as well. That is the reason for the requirement of having a trusted commodity server, which does not collude with any vertical server. Despite this restriction, using a commodity server improves the runtime effectively, and – according to [10] – finding such a *cs* is feasible in practice. It makes no difference if the *cs* is semi-honest (also known as honest-but-curious) or malicious, because it acts only as a receiver in the training mode and receives only unconditional messages that it cannot manipulate by own messages in the classification mode.

– **Disclosure of the input of other parties:** In the multi-party setting, there is a general risk of collusion between the data holding parties to combine their input data maliciously in order to violate an individual's privacy. However, this risk exists independent of the data mining protocols, hence it cannot be prevented in their design. As a protocol dependent aspect, we consider a case where b colluding vertical servers try to reveal the input data of one or more other servers. Looking into the Algorithm 1 indicates that there is no difference between semi-honest and malicious behaviour again. In the first part of algorithm (lines 3–8), every vertical server distributes numbers independent of each other. In the second part (lines 9–12), $S_i[j]$ comprises either $r_i^0[j]$ or another random number but no direct input data besides the numbers of other parties. This procedure happens independent of the messages of other parties, and consequently the public input of any party ($S_i[j]$) does not reveal any information if the numbers sent by other parties are generated randomly or with a malicious intention.

One adversary might try to find out whether an element d is supported by all parties or a particular vertical server p_x. The question if all parties hold d is defined by $\sum_{i=0}^{n}(R_i - S_i)$. Since each $S_i[j]$ is directly sent to the commodity server, only the commodity server (or all p_i together) is able to learn it. In order to find out whether a particular vertical server p_x holds d, adversaries have to know if:

$$S_x = \sum_{t=0}^{T} r_{x-t}^t \mod n \Leftrightarrow S_x = \sum_{t=1}^{T} r_{x-t}^t \mod n + r_x^0. \tag{2}$$

The only exception is if all p_i support d, because in that case, it is trivial that a particular vertical server does it too. S_x is only known by p_x and the commodity server. Given a random S_x, it cannot be calculated from other values. Arranging the vertical servers in a cycle in clockwise direction, $\sum_{t=1}^{T} r_{x-t}^t \mod n$ can only be calculated by the T servers on the right side of

p_x. r_x^0 is only known by p_x or can be calculated from $R_x - \sum_{t=1}^{T} r_x^t$, where R_x is only known by p_x and commodity server and $\sum_{t=1}^{T} r_i^t$ can only be calculated by the T servers on the left side of p_i. In conclusion, at least $\min(n - 1, 2T)$ colluding vertical servers and the commodity server are necessary to find out whether a particular party p_x supports an element d.

4 Experimental Results and Complexity

4.1 Experiment Settings and Datasets

We implemented and tested the main random forest framework in Java, with four versions of the private set intersection:

1. The CVSSI protocol as designed in Algorithm 1.
2. Du02 version where we used a modified version of the scalar product protocol by Du and Zhan [10], such that the commodity server receives the output. This version has the constraint of a commodity server like our CVSSI protocol and is very fast, but can only be used for two-party problems.
3. A simple asymmetric public-key encryption scheme (Paillier encryption) that Vaidya and Clifton first used in the distributed decision tree context [8, 26], because it fulfills the requirement of additive homomorphy. One party encrypts the identifier of d if it supports the element d else a zero with the public key. Then, each vertical server multiplies the encrypted intermediate result of its predecessor with a one if it supports d and with a zero if it does not. At the end, the results for each d are summed up, so that the total result is the encryption of d, because in our case only one d is supported by all parties together. Only the commodity server has the private key and can decrypt the result of the last vertical server and gets either a one or a zero. For simplicity, we do not use the state-of-the-art Paillier encryption, but give an idea of homomorphic encryption techniques.
4. A procedure with public splits like by Suthampan and Maneewongvatana [23] instead of a private set intersection method as a baseline, which is very straightforward, but reveals information we want to protect.

All experiments were executed on a single device with a dual core intel i7-5500U cpu and a 8 GB RAM. For the current results, we did not use a framework to simulate bandwidth and latency of a network of different devices. We tested the scalability on different real-world datasets of the UCI Machine Learning Repository with different parameters: number of vertical servers n, the collusion threshold parameter T, and the number of leaf nodes. The number of leaf nodes is β^δ, if the tree depth δ and the number of splits in a branch β are fixed in our experiments.

Table 2. Communication costs

Intersection protocol	2 vertical server	>2 vertical server
Du02	$4\,s * b$	–
CVSSI	$6\,s * b$	$s * n * (T + 2) * b$
Paillier encryption	$2\,s * B$	$s * n * B$
Non private	–	–

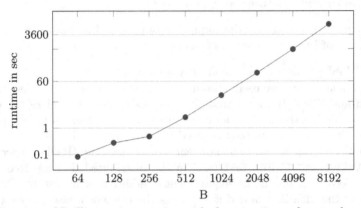

Fig. 2. Runtime of Paillier set intersection with three parties and vector length 1000.

4.2 Complexity Comparison

The computational complexity of Algorithm 1 is $O(s \times n \times T)$, with the vector length s, the number of vertical servers n and $T < n$. The algorithm Du02 requires $O(s)$ elementary operations, which is of the same order in a two-party setting. The Paillier encryption version consumes $(s+1)n$ bit multiplications for the encryption and summation and one bit exponentiation for the decryption. The fourth, not private version does not use a set intersection protocol. Instead, one responsible party broadcasts the supported records at each branch and leaf node. Hence, there are no computation and communication costs for a set intersection computation. Table 2 shows the communication costs depending on the bit length b of a data type and B as the bit length of a public key.

Apart from some initialization costs, Algorithm 2 calls $m \times o$ times a private set intersection protocol with the input size β^δ. Before the set intersection can be executed, every vertical server has to do the preprocessing step of filling the input vectors with a complexity of $O(\beta^\delta)$. This equates to a total computation and communication complexity of $O(m \times n \times o \times T \times \beta^\delta)$ in connection with Algorithm 1 (CVSSI). The total complexity of the non-private version is $O(m \times n \times o \times \beta^\delta)$. This is because all supported records are broadcast to every vertical server at each tree node once. In the deeper nodes, the number of supported instances is much smaller than m.

4.3   Performance Analysis

Figure 2 visualizes the exponential dependency of the runtime of homomorphic encryption schemes in relation to key bit length B. The German federal office for information security recommends a key length of 2,000 to 3,000 bits [7], which leads to a runtime of a few minutes for a single small vector with 1,000 elements in [26] and our experiments. This is rather infeasible for the whole tree building and classification procedure. Our CVSSI algorithm requires 1 ms for this task. The recently published private set intersection protocol by Kolesnikov *et al.* [17] runs also in less than a second in their environment and might be an alternative building block. However, one has to consider that the authors in [17] used a much more elaborate framework to simulate communication costs than we did.

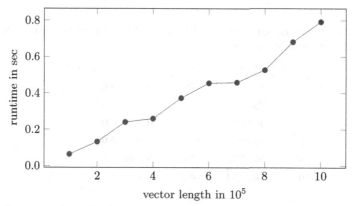

Fig. 3. Runtime of CVSSI (number of vertical servers: 4, T = 3, 100 repetitions).

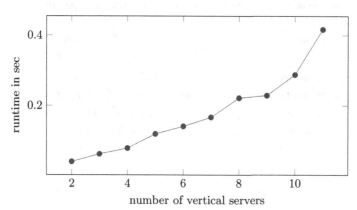

Fig. 4. Runtime of CVSSI (vector length: 10^6, $T = 3$).

Figures 3 and 4 confirm the linear scalability of our CVSSI algorithm in terms of n and o. We obtain similar results for T. They also show that the protocol is feasible for larger vector sizes and a higher number of involved servers. Table 3 contrasts the runtimes of the four versions on the small car dataset (1728 instances) with a small number of trees (5) in a two vertical server setting. 90% of the instances are used for training and 10% for testing. As expected, the Paillier encryption version requires several minutes, although we set B to the unacceptably low value of 64. Our approach, the CVSSI protocol, takes less than half a second per tree. The Du02 version performs up to six times faster than our approach, which might be due to more effective vector operations in our implementations. The variant without private set intersection suggests further potential for improvement, but suffers from the inference problem. Table 4 shows the runtime of CVSSIP on real-world data sets with five parties and 20 randomly generated trees in seconds. This suggests that the approach is feasible in practice.

Table 3. Runtime on the car dataset, with two parties, 4 splits per branch, tree depth 5, in seconds and 10 random trees.

Trees	Paillier enc.	CVSSIP	Du02	Without PSI
1	115.519	0.464	0.080	0.003
2	238.991	0.922	0.169	0.006
4	461.899	1.796	0.321	0.008
6	693.079	2.701	0.466	0.011
8	919.099	3.591	0.639	0.014
10	1,165.374	4.631	0.833	0.015

Table 4. Runtime of CVSSIP on UCI ML data sets with five parties and 20 random trees.

Dataset	n	β^δ	runtime (s)
Cars	1,728	4^6	155.7
Contraceptive	1,473	2^9	16.8
Hepatitis (no missing values)	80	2^{17}	11,507.4
Nursery	12,960	5^7	22,783.6
Phishing websites	11,055	2^{13}	16,231.7
Thoraric surgery	470	2^{13}	173.1

5   Conclusion

We presented a new architecture which is a hybrid approach of private evaluation and classification on vertically partitioned data. This setting might become more interesting in the future with the increasing use of private data and collaborations of companies, governments and different organizations. We provided a closed, lightweight and feasible solution with adaptable security levels. Additionally, it is highly parallelizable. The main drawback is the assumption of a central non-colluding commodity server. Making use of the results of Kolesnikov *et al.* on 1-out-of-n oblivious transfer and private set intersection [16,17], it may be possible to overcome this dependency in the future.

References

1. Aggarwal, C.C., Yu, P.S.: A general survey of privacy-preserving data mining models and algorithms. In: Aggarwal, C.C., Yu, P.S. (eds.) Privacy-Preserving Data Mining, pp. 11–52. Springer, Boston (2008). https://doi.org/10.1007/978-0-387-70992-5_2
2. Blockeel, H., Raedt, L.D.: Top-down induction of first-order logical decision trees. Artif. Intell. **101**(1), 285–297 (1998)
3. Bojarski, M., Choromanska, A., Choromanski, K., Lecun, Y.: Differentially- and non-differentially-private random decision trees, October 2014
4. Bost, R., Popa, R.A., Tu, S., Goldwasser, S.: Machine learning classification over encrypted data. Cryptology ePrint Archive, Report 2014/331 (2014)
5. Breiman, L.: Random forests. Mach. Learn. **45**(1), 5–32 (2001)
6. Brickell, J., Porter, D.E., Shmatikov, V., Witchel, E.: Privacy-preserving remote diagnostics. In: Proceedings of the 14th ACM Conference on Computer and Communications Security (CCS), pp. 498–507 (2007)
7. Bundesamt für Sicherheit in der Informationstechnik: Kryptographische verfahren: Empfehlungen und schluessellaengen, May 2018. https://www.bsi.bund.de/SharedDocs/Downloads/DE/BSI/Publikationen/TechnischeRichtlinien/TR02102/BSI-TR-02102.pdf?__blob=publicationFile&v=8
8. Damgård, I., Jurik, M., Nielsen, J.: A generalization of paillier's public-key system with applications to electronic voting. Int. J. Inf. Secur. **9**, 371–385 (2003)
9. Drummond, C., Holte, R.C.: Exploiting the cost (in)sensitivity of decision tree splitting criteria. In: International Conference on Machine Learning (ICML), pp. 239–246 (2000)
10. Du, W., Zhan, Z.: Building decision tree classifier on private data. In: Proceedings of the IEEE International Conference on Privacy, Security and Data Mining (CRPIT)-Volume 14, pp. 1–8 (2002)
11. Emekci, F., Sahin, O., Agrawal, D., Abbadi, A.E.: Privacy preserving decision tree learning over multiple parties. Data Knowl. Eng. **63**(2), 348–361 (2007)
12. Friedman, A., Schuster, A.: Data mining with differential privacy. In: Proceedings of the 16th ACM SIGKDD International Conference on Knowledge Discovery and Data Mining (KDD), pp. 493–502 (2010)
13. Wu, D.J., Feng, T., Naehrig, M., Lauter, K.: Privately evaluating decision trees and random forests. Proc. Priv. Enhancing Technol. **2016**, 335–355 (2016)
14. Jagannathan, G., Pillaipakkamnatt, K., Wright, R.N.: A practical differentially private random decision tree classifier. In: IEEE International Conference on Data Mining Workshops, pp. 114–121, December 2009
15. Kaghazgaran, P., Takabi, H.: Differentially private decision tree learning from distributed data, May 2015

16. Kolesnikov, V., Kumaresan, R., Rosulek, M., Trieu, N.: Efficient batched oblivious PRF with applications to private set intersection. In: Proceedings of the ACM SIG SAC Conference on Computer and Communications Security (CCS), pp. 818–829 (2016)
17. Kolesnikov, V., Matania, N., Pinkas, B., Rosulek, M., Trieu, N.: Practical multi-party private set intersection from symmetric-key techniques. In: Proceedings of the ACM SIG SAC Conference on Computer and Communications Security (CCS), pp. 1257–1272 (2017)
18. Lindell, P.: Privacy preserving data mining. J. Cryptol. **15**(3), 177–206 (2002)
19. Liu, X., Li, Q., Li, T., Chen, D.: Differentially private classification with decision tree ensemble. Appl. Soft Comput. **62**, 807–816 (2018)
20. Ma, Q., Deng, P.: Secure multi-party protocols for privacy preserving data mining. In: Li, Y., Huynh, D.T., Das, S.K., Du, D.-Z. (eds.) WASA 2008. LNCS, vol. 5258, pp. 526–537. Springer, Heidelberg (2008). https://doi.org/10.1007/978-3-540-88582-5_49
21. Quinlan, J.: Induction of decision trees. Mach. Learn. **1**(1), 81–106 (1986)
22. Sravya, C.L., Lakshmi, G.R.: Privacy-preserving data mining with random decision tree framework. IOSR J. Comput. Eng. **19**, 43–49 (2017)
23. Suthampan, E., Maneewongvatana, S.: Privacy preserving decision tree in multi party environment. In: Lee, G.G., Yamada, A., Meng, H., Myaeng, S.H. (eds.) AIRS 2005. LNCS, vol. 3689, pp. 727–732. Springer, Heidelberg (2005). https://doi.org/10.1007/11562382_75
24. Tai, R.K.H., Ma, J.P.K., Zhao, Y., Chow, S.S.M.: Privacy-preserving decision trees evaluation via linear functions. In: Foley, S.N., Gollmann, D., Snekkenes, E. (eds.) ESORICS 2017. LNCS, vol. 10493, pp. 494–512. Springer, Cham (2017). https://doi.org/10.1007/978-3-319-66399-9_27
25. Teng, Z., Du, W.: A hybrid multi-group privacy-preserving approach for building decision trees. In: Zhou, Z.-H., Li, H., Yang, Q. (eds.) PAKDD 2007. LNCS (LNAI), vol. 4426, pp. 296–307. Springer, Heidelberg (2007). https://doi.org/10.1007/978-3-540-71701-0_30
26. Vaidya, J., Clifton, C.: Privacy-preserving decision trees over vertically partitioned data. In: Jajodia, S., Wijesekera, D. (eds.) DBSec 2005. LNCS, vol. 3654, pp. 139–152. Springer, Heidelberg (2005). https://doi.org/10.1007/11535706_11
27. Zhan, J.Z., Chang, L.W., Matwin, S.: Privacy-preserving multi-party decision tree induction. In: Farkas, C., Samarati, P. (eds.) DBSec 2004. IIFIP, vol. 144, pp. 341–355. Springer, Boston, MA (2004). https://doi.org/10.1007/1-4020-8128-6_23

Privacy Risk for Individual Basket Patterns

Roberto Pellungrini[1]([✉]), Anna Monreale[1], and Riccardo Guidotti[1,2]

[1] University of Pisa, Largo B. Pontecorvo, 3, Pisa, Italy
{roberto.pellungrini,anna.monreale,riccardo.guidotti}@di.unipi.it
[2] KDDLab, ISTI-CNR, Via G. Moruzzi, 1, Pisa, Italy
riccardo.guidotti@isti.cnr.it

Abstract. Retail data are of fundamental importance for businesses and enterprises that want to understand the purchasing behaviour of their customers. Such data is also useful to develop analytical services and for marketing purposes, often based on individual purchasing patterns. However, retail data and extracted models may also provide very sensitive information to possible malicious third parties. Therefore, in this paper we propose a methodology for empirically assessing privacy risk in the releasing of individual purchasing data. The experiments on real-world retail data show that although individual patterns describe a summary of the customer activity, they may be successful used for the customer re-identifiation.

1 Introduction

Retail data are one of the most important source of information that enables commercial companies in understanding their customers behavior by analyzing their purchasing patterns. In the literature, many data mining methods have been proposed to extract customer patterns describing frequent itemsets [2], top-k frequent itemsets [29], regular itemsets [14]. All these individual purchasing models may enable not only the understanding of collective and individual behaviors, but also the development of data-driven services such as personal recommendation systems able to capture the customers' preferences.

Unfortunately, the analysis of retail data might lead to the inference of highly sensitive information about individuals. Thus, in the literature some works have addressed the problem of privacy issues in market basket data. Some of them proposed a methodology for the empirical privacy risk evaluation [20], while others proposed some approaches for guaranteeing privacy protection [15,30]. However, all these works are focused on the study of the privacy issues in the released purchasing data, that is, they study the potential privacy risk related to the release of raw data collected from individuals. Instead, in this paper we propose to study the privacy risk assessment of individual purchasing models extracted from the purchasing data of individuals during analysis processes. Specifically, we identify two types of individual purchasing models: individual

C. Alzate et al. (Eds.): MIDAS 2018/PAP 2018, LNAI 11054, pp. 141–155, 2019.
https://doi.org/10.1007/978-3-030-13463-1_11

models composed by a single pattern and individual models composed by a set of patterns. Then, we define the privacy attack models and the methods for their simulation. Finally, we simulate these attacks on real-world retail data and we analyze the privacy risk distributions trying also to identify the properties of bought items that can lead to customer re-identification by her patterns. The results show that, although individual patterns are models that abstract from the details of the raw data, they are able to capture peculiarities of the customer behavior which often lead to the customer re-identification.

The rest of the paper is organized as follows. In Sect. 2, we discuss the related work. Section 3 introduces the data models used for representing retail data. In Sect. 4, we define the privacy risk assessment methodology including the privacy attacks. Section 5 shows the results of our experiments and, finally, Sect. 6 concludes the paper.

2 Related Work

Customer profiling is a process widely used in economy since long time ago for direct marketing, site selection, and customer relationship management. The process of construction and extraction of a personal data model formed by personal patterns is generally referred to as *user profiling*. A user profile contains the systematic behaviors expressing the repetition of habitual actions, i.e., personal patterns. These patterns can be expressed as simple or complex indexes [10], behavioral rules [14], set of events [13], typical actions [28], etc. Profiles can be classified as individual or collective according to the subject they refer to [9,16]. An *individual* profile is built considering the data of a single person. This kind of profiling is used to discover the particular characteristics of a certain individual, to enable unique identification for the provision of personalized services. We talk about *collective* data models when personal data or individual models generated by individual profiling are aggregated without distinguishing the individuals.

With respect to market basket analysis, customer profiling can play today a very important role. Nowadays the market is characterized by being global, products and services are almost identical and there is an abundance of suppliers. Therefore, instead of targeting all the customers equally, a company can select only those customers who meet certain profitability criteria based on their individual needs and buying patterns [4]. To achieve this goal, the customers must be described by characteristics valuable for the business, like the demographic ones, the lifestyle, and the shopping habits. These targets can be reached through customer profiling. By knowing the profile of each customer, a company can treat a customer according to her individual needs and increase the lifetime value of the customer [4]. Furthermore, customer profiling is a key element which impacts into the decisions in product life cycle cost [7]. One of the first methodology proposed to analyzed shopping session is [3] where frequent patter mining rules are defined. In [1] is described a system exploiting these rules for building personal profiles on transactional histories. The profiles consists of a set of rules describing customers' behavior. However, this system requires a constant user feedback to assess the pattern validity and parameter setting. An automatic and parameter-free approach to derive personal patterns is proposed in [13]. An evolution of [13] that also consider the temporal dimension is described in [14]. In [31] the authors analyze customers' shopping behaviors with respect to both on product profiles and customer profiles. The product profile is characterized by a set of features describing the product. The customer profile this time is an index expressing the level of interest in product features calculated using the product profiles. A two-stage clustering technique is used to find the group of customers that have similar interests and then extract rules from each cluster. In [10] the authors propose two indexes that consider the level of repetitiveness in both the basket composition and also in the temporal and spatial dimension of shopping purchases, i.e., when and where the customers go to the supermarket. Other forms of customer profiling on market basket data like those described in [11,12] adopt ad vector based modeling.

In existing literature, the privacy risk for the sharing of retail data or customer's profiles is not considered. This is especially interesting considering the high amount of privacy related literature.

A vastly used privacy-preserving model, and one of the models of our choosing for this paper, is k-anonymity [23], which requires that an individual should not be identifiable from a group of size smaller than k based on a subset of her own attributes used to univocally identify her, called quasi-identifiers. In [5] the authors present a set of attacks on the k-anonymity model to prove it's possible weaknesses while in [34] a graph-attack method based on k-anonymity to defend from possible privacy attacks is proposed. More recently, in [19] the k-anonymity model has been used as a base to propose a privacy framework for the systematic simulation of privacy attacks, then applied to mobility data. For retail data very little has been done in terms of privacy risk assessment. In [21] authors propose a framework for anonymizing transactional data, and in [33] and [32] the authors propose various methods for privacy preserving data publishing with transactional and retail data.

For privacy risk assessment, a fundamental work is the LINDDUN methodology, presented in [6]. The LINDDUN framework for privacy threats analysis is largely based on the privacy threat modeling framework STRIDE [25] used in software-based systems. Other methods for privacy risk evaluation have been published recently such as in [27], where the authors elaborate an entropy-based method to evaluate the disclosure risk of personal data, trying to manage quantitatively privacy risks.

In this paper we use a well known technique to match records of different data-set known as distance based record linkage. This technique was first introduced in [17], and allows for the matching of records from different data-sets based on a measure of distance between records. Records that have minimal distance between each other are considered to belong to the same individual and are matched. Different variations of this technique have been used in privacy literature such as in [26], where the Mahalanobis distance is used for distance based record linkage.

3   Retail Data

Retail data is generally collected through membership programs: customers who wish to do so, voluntarily agree to such programs in order to receive some benefits through the use of a specific membership card, the data about their purchases is subsequently collected. The raw data of each individual is represented by baskets. A basket is a set of items purchased by the individual during a shopping session. We consider baskets with no repetitions, i.e., proper sets where items can appear only once. Therefore, and individual may have multiple baskets associated to her.

Definition 1 (Basket). *We define a basket (or transactions) b as a subset of items such that $\emptyset \subset b_i \subseteq I$ where $I = \{i_1, \ldots, i_D\}$ is the set of all D items.*

Definition 2 (Basket History). *We define the basket history $B_u = \{b_1, \ldots, b_N\}$ as the set of N baskets (or transactions) belonging to the individual u.*

Such data is usually used to perform analysis of various kind, from association rule mining [2] to clustering [8]. In this paper we focus on transactional clustering, as performed with the state-of-the-art algorithm TX-Means [13]. TX-Means is a parameter-free clustering method that follows a clustering strategy similar to TX-Means [18] designed for finding clusters in the specific context of transactional data. TX-Means automatically estimates the number of clusters and it also provides the *representative basket* of each cluster, which summarizes the pattern captured by that cluster. The representative baskets correspond to the centroids of the sub-clusters and are calculated adopting the procedure described in [8]. Therefore, the output of TX-means, consisting in the representative baskets, is a set of typical patterns that represent recurring purchasing behavior of each individual. Note that, TX-means is only one of the algorithms able to discover purchasing patterns. We point out that different algorithms may discover purchasing patterns capturing different properties. For example, a standard pattern mining algorithm as Apriori [2] is able to extract frequent patterns that differ from recurrent patterns. However, it requires the minimum support as parameter that, from a personal data analytics perspective [9], should be personally tuned of each user. Another example of pattern can represent the top-k frequent items. However, in all these cases a *pattern* may be modelled similarly to a set of baskets.

Definition 3 (Patterns). *We define as $P_u = \{p_1, p_2, \ldots, p_M\}$ the sets of patterns of the individual u, where each $p_i \subseteq I$ and I is the set of all D items.*

4   Privacy Risk Assessment Methodology

In literature there are several notable methodologies proposed to assess privacy risks. The definition of privacy that we use was first introduced in [23]. To assess privacy risk we adopt the framework proposed in [22] that is also used in [19].

The basic assumption is that a malicious third party, commonly referred to as the *adversary*, gathers some background knowledge about an individual, i.e., a subset of the information related to the individual. Then, the adversary tries to re-identify the individual in a published data-set using that background knowledge. If successful, the adversary could then be able to retrieve the complete information associated to the individual, i.e., the adversary could gain access to all the records regarding the individual. Thus, the general approach in applying this framework is to first determine the possible background knowledge of an adversary, then simulate an attack on the data using such background knowledge, empirically compute the privacy risk, and finally explore and analyze the results to assess privacy risk.

In order to understand the nature of privacy risk in retail data we define a set of attacks based upon the above framework to explore the privacy risk in this kind of data.

Patterns Against Patterns. In the first attack we consider an adversary who tries to understand how unique the individual patterns extracted by clustering algorithms are. To this end, we conducted our study on two types of individual purchasing patterns, extracted by using two different clustering algorithms. The first one is a very simple baseline approach that for each individual u extracts a single pattern consisting in the set of her most frequent k items. In other words, for each individual u we have only one pattern in P_u, i.e., $p = \{i_1, i_2, \ldots, i_k\}$. In the rest of the paper we refer to this patterns as simple patterns. The second approach is the state-of-the-art clustering algorithm, TX-Means [13]. Using this more complex approach every customer can be characterized by a different number of patterns. Every pattern $p_j \in P_u$ corresponds to a *representative basket* extracted by TX-Means. In the rest of the paper we refer to this patterns as TX-means patterns. A representative basket is a virtual transaction that approximates a set of similar baskets, therefore capturing the items that best characterize it, i.e., the typical combination of items expected to appear in any of its baskets. Then, we define an attack where an adversary gathers a certain number of the patterns for each individual and tries to re-identify the individual in the whole set of published patterns.

For the first approach, the privacy risk of an individual is given by the number of other individuals sharing the same pattern.

Definition 4 (Single Pattern Risk). *Given an individual u with a single pattern in P_u, we define her privacy risk as: $Risk_u = \frac{1}{|M_{P_u}|}$, where $|M_{P_u}|$ is the cardinality of the set of individuals having the same pattern in P_u. This measure ranges from 0 to 1.*

For the second approach, where multiple patterns belong to the same individual, we relied on a systematic exploration of all the possible background knowledge of a certain length h. For instance, if a customer has 3 patterns $\{p_1, p_2, p_3\}$ and we assume an adversary knows 2 of them, we calculate the privacy risk exploring all the possible combinations of the 3 patterns with length 2. In the above example, the following three background knowledge would be used: *(i)*

$\{p_1, p_2\}$, (ii) $\{p_1, p_3\}$, (iii) $\{p_2, p_3\}$. Each combination is compared with all the patterns in the published dataset, i.e., we check how many customers have the same patterns in the data.

Definition 5 (Multiple Patterns Risk). *Let u an individual with multiple patterns in P_u and let C_h be the set of possible combinations of patterns with length h. The customer privacy risk is defined as: $Risk_u = \frac{1}{\min_c(|M_c|)}$, where M_c is the set of customers having a particular combination of patterns $c \in C_h$. This measure ranges from 0 to 1.*

This is a worst-case based approach, as we use the most unique patterns to calculate the risk, given by the use of minimum value of $|M_c|$.

Patterns Against Baskets. In the definition of the second attack we assume that an adversary might get access to the patterns dataset $\mathcal{P} = \{P_{u_1}, \dots P_{u_U}\}$ and use it to attack the basket history data $\mathcal{B} = \{B_{u_1}, \dots B_{u_U}\}$, where U is the number of different customers. This could happen for example in the case when the patterns are publicly made available because considered safe, and the adversary gets access to the anonymized basket history data. In this case, we cannot directly compare the pattern of an individual with the customer baskets to find a match, but we need to identify the possible basket history $B_i \in \mathcal{B}$ that could have generated the known pattern $P_i \in \mathcal{P}$. Thus, we should link the different basket histories in $\mathcal{B}$ with each pattern in $\mathcal{P}$ through the use of a distance measure. In particular, we propose to use the distance function introduced in [17]. The adversary will match each pattern in $\mathcal{P}$ with the closest basket history in $\mathcal{B}$. Clearly, if the distance between the pattern of the customer u in $\mathcal{P}$ and the basket history of u is the minimum, then the two records of that customer are correctly matched.

We recall that the set of the representative patterns of each individual is computed with either TX-means or the baseline approach. To calculate the distance between this the records in the data to be matched we propose to use a modified version of the Jaccard distance.

Definition 6 (Jaccard Distance). *Let A and B be two sets. The Jaccard distance is defined as: $J(A, B) = \frac{|A \cap B|}{|A \cup B|}$.*

Definition 7 (Minimum Jaccard). *Let A and $Y = \langle b_1, b_2, \dots, b_m \rangle$ be a set and a set of sets respectively. The Minimum Jaccard distance is defined as: $MJ(A, Y) = \min_{i=1,2,\dots,m}(J(A, b_i))$.*

Definition 8 (Best Jaccard). *Let $X = \langle a_1, a_2, \dots, a_n \rangle$ and $Y = \langle b_1, b_2, \dots, b_m \rangle$ be two set of sets, with $n \leq m$. The Best Jaccard distance is defined as: $BJ(A, Y) = \sum_{i=1}^{n} MJ(a_i, Y)$.*

Using the Best Jaccard distance, we can calculate the number of correct matches that an adversary could make using the pattern dataset to attack the basket history dataset. Now, we are ready to introduce the definition of the privacy risk in this particular setting.

Definition 9 (Patterns Against Baskets Risk). *Let U be the set of all individuals and M be the set of individuals for whom $BJ(P_u, B_u)$ has the minimum value. Then, we define the privacy of the dataset as: $Risk = \frac{|M|}{|U|}$. This measure ranges from 0 to 1.*

This approach dates back to [24]. Note that, in this case, we cannot directly express a measure for individual risk, since an adversary either correctly matches two records of the same individual or doesn't.

5 Experiments

We performed experiments on real world dataset provided by UniCoop Tirreno, a large Italian supermarket chain. Customers are provided with a loyalty card which allows to link different shopping sessions, and therefore reconstruct their personal shopping history. We analyzed a dataset of 2,021,414 shopping sessions, i.e., baskets, performed by 8564 individuals between the 2010 and 2012 in Leghorn province. These customers are "loyal customers", i.e., customers active in at least ten months every year. For each customer we have on average 240 baskets, containing 100 different items, and the average basket length is 8 items.

For each customer we extracted her typical patterns using the two approaches discussed previously in Sect. 4. Using the baseline approach for the patterns extraction, we obtained patterns considering the k-most frequent items for each person, with k ranging from 1 to 10. Applying TX-Means we extracted a total of 38,068 patterns, more than 4 patterns per individual on average.

5.1 Patterns Against Patterns

In this section we analyze the empirical results related to the privacy risk for the patterns against patterns attack.

Simple Patterns Against Simple Patterns Risk. The first experiment that we performed is the simulation of the patterns against patterns attack using simple patterns, i.e., the top k items by frequency for each individual.

In Fig. 1 we show the distribution of privacy risk for this attack using the baseline approach, by increasing the value of k, i.e., increasing the number of items in the k-most frequent patterns. We observe that, with 2 items (Fig. 1(a)), we have a lower distribution of the privacy risk. But increasing the number of known items, the level of risk increases rapidly. With 4 items (Fig. 1(c)), more than half of the population shows risk 1, i.e. maximum risk. Beyond $k = 5$ the risk becomes 1 for more than 95% of the population. Starting from the different top-k items of each individual for any value of k, we analyzed the length of the shortest simple pattern of each individual that yields privacy risk 1. The idea is to understand for the customers the distribution of risky k values.

Figure 2 reports the result of this analysis. We found a rather classical Gaussian distribution, with a peak around 4 as expected. Moreover, we also tried to characterized the risky top-k items. To this end, for each customer we selected

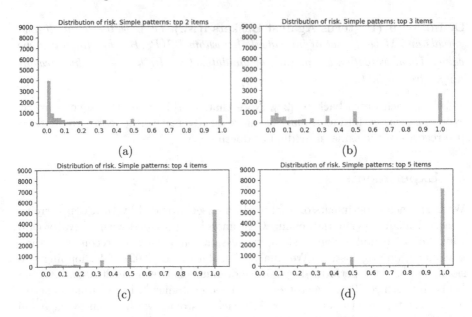

Fig. 1. Distribution of risk in simple patterns attack

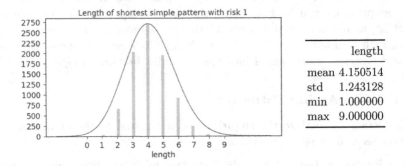

Fig. 2. Distribution of the length of the shortest simple patterns that yield risk 1

the shortest pattern that yield risk 1 and among the item composing them we identify those having the lowest global frequency in the basket history data and the lowest frequency in the set of top-k patterns. In practice, these items are bought by very few customers but are very frequent in the basket history of their customers. Given this property they probably are the cause of the customer high privacy risk. In Table 1 we report the list of the 10 items with lowest global frequency that appear in a low number of simple patterns. We observe that they are very particular items and most of them are not food items.

Table 1. Infrequent items within the simple patterns

Macro-sector	Category
No food	Deodorants for environments
Grocery	Honey
No food	Hardware
No food	Anniversary card
Fresh food	Fruit beverages
No food	Woman's socks
No food	Sandpaper
Fresh food	Sheep meat
No food	Flowers
No food	Chemical products

TX-means Patterns Against TX-means Patterns Risk. The second experiment is focused on the simulation of a patterns-against-patterns attack using the individual models extracted with the TX-means algorithm. Each individual is hence represented by multiple patterns. To compute the privacy risk we checked all possible combinations of patterns of length h, with h values ranging from 1 to 3. We report the results in Fig. 3. We can see that changing the value of h does not impact on the level of risk as with just one pattern (Fig. 3(a)), it is possible to correctly re-identify more than 99% of the individuals. This means that almost every individual has at least one unique pattern that represents him. This is not surprising, since TX-means is an advanced algorithm for personal data analytics and yields highly personalized results. We can further explore the results by looking at the length of the patterns and the privacy risk that they yield.

Figure 4 highlights that there is no clear correlation between privacy risk values and pattern length. However, we observe that there is no pattern with length greater than 5 that yield a risk lower than 1. As for simple patterns, this suggests that longer and more complex patterns are more unique and personal; as a consequence, they lead to the identification of the individuals. For the TX-means patterns we performed the same analysis already presented for simple patterns; in other words, we analyzed the distribution of the length of the shortest pattern that for each individual yields the maximum risk.

We observe that TX-means provides longer patterns on average and the distribution presents a typical long tail shape. In Table 2 we report the list of the 10 items with lowest global frequency that appear in a low number of TX-means patterns. As for the simple patterns, we highlight that most of them are not food items but their categories are more common with respect to the simple patterns. Overall these experiments suggests that representative patterns extracted with either naive or advanced techniques are inherently unique. An individual may be easily re-identifiable using these patterns even with a small number of items.

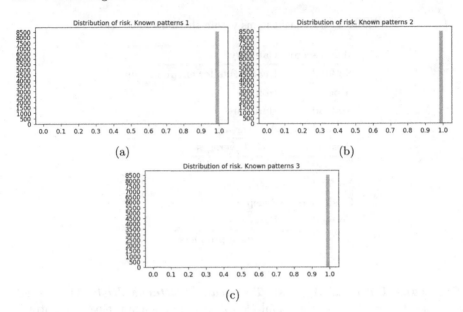

Fig. 3. Distribution of risk in TX-means patterns attack

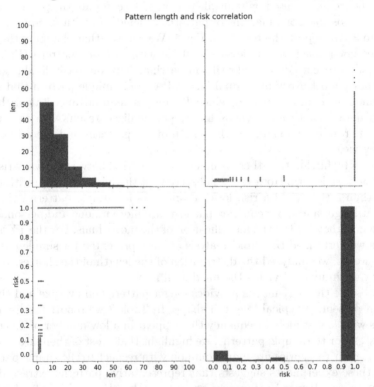

Fig. 4. Correlation between pattern length and privacy risk

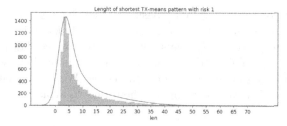

	length
mean	9.046007
std	8.473491
min	1.000000
max	71.000000

Fig. 5. Distribution of the length of the shortest TX-means patterns that yield maximum risk

Table 2. Infrequent items within the TX-means patterns

Macro-sector	Category
No food	Manual tools
Fresh food	Frozen meat
Fresh food	Poultry for birds and rabbits
Fresh food	Milk
No food	Christmas decoration
No food	Underwater gear
No food	Electrical equipment
No food	House carpets
No food	House decoration
No food	Glasses

As for the items themselves we see a fairly broad characterization, however, we can conclude that non-food related items are much more distinctive and may lead to higher chances of re-identification (Table 2).

5.2   Patterns Against Baskets

In this section we analyze the empirical privacy risk in case of the patterns against baskets attack.

Simple Patterns Against Baskets. The first experiment is based on the simulation of a patterns against baskets attack using simple patterns. We recall that for this attack risk is evaluated globally for the entire data-set and not individually. We performed distance based record linkage with simple patterns of 2, 4 and 5 items. For simple patterns of length 2 we have only 27 correct matches out of the total population of 8,564 customers. This yields a risk of 0.003. For simple patterns of length 4 we have 298 correct matches, yielding a risk of 0.034. For patterns of length 5 we have 388 correct matches, yielding a risk of 0.045. These low values are probably due to several factors: while we have shown previously that simple patterns are quite unique, they are not particularly representative of the individual's baskets. Also, having only one pattern significantly diminishes the information used for the linkage. Because of how we compute distance, having only one simple pattern implies that such distance fall in the range 0 to 1. This leads to a high number of individuals with minimum distance, therefore impeding a univocal matching. We can conclude that simple patterns pose a relatively low threat when used to attack the raw data.

TX-means Patterns Against Baskets. The second experiment is based on the simulation of a patterns against baskets attack using the patterns extracted with the TX-means clustering algorithm. As for the previous case, the risk is calculated for the entire data-set. With the TX-means patterns we have that 5,781 individuals out of the total population of 8,564 customers are correctly matched, i.e., the distance between the TX-means patterns of those individuals and their basket data is minimal. This yields a risk of 0.675. We can now characterize the individuals correctly matched, by looking at their patterns and baskets.

Table 3. Characterization of matched individuals in the TX-means patterns against baskets attack

	Patterns: std of length	Patterns: mean length	Number of patterns	Number of baskets	Baskets: std of length	Baskets: mean length
Mean	4.811004	13.049558	4.820446	244.230064	6.002396	10.897940
Std	3.996948	7.899513	3.453788	201.790281	2.873166	5.264362
Min	0.000000	2.200000	1.000000	10.000000	0.708363	1.744063
Max	26.051631	71.000000	25.000000	1646.000000	26.411782	43.282051

Table 4. Characterization of non matched individuals in the TX-means patterns against baskets attack

	Patterns: std of length	Patterns: mean length	Number of patterns	Number of baskets	Baskets: std of length	Baskets: mean length
Mean	2.884773	10.653819	3.665469	219.015451	4.884745	8.000338
Std	3.385043	7.122223	3.735964	220.721776	2.372840	3.951333
Min	0.000000	1.000000	1.000000	10.000000	0.535428	1.221429
Max	25.500000	53.000000	26.000000	2025.000000	16.146130	31.976744

In Tables 3 and 4 we gathered some statistics for the individuals correctly matched and those who were not matched. For each individual, we gathered the mean length of her patterns and her baskets as well as the standard deviation for such lengths and the number of patterns and baskets. In the tables we show mean, standard deviation, min value and max value for the aforementioned measures. If we compare the statistics in the two table we can see that there are not many differences. However, we observe that, for the individuals that were not re-identified by the attack, we have fewer, shorter patterns and baskets on average, again, confirming that higher risk is related to lengthier baskets and/or patterns.

6   Conclusion

In this paper we have studied the privacy risk assessment of individual purchasing patterns. In the study we have taken into consideration two different individual patterns: the top-k items of an individual and the representative patterns extracted by TX-means. After defining, two possible attacks that exploit individual patterns for customers re-identification, we have performed their simulation on real-world data. The empirical results on the privacy risk distributions show that individual patterns often lead to the re-identification of most of the customers because they accurately describe some customer habits that make him unique. This preliminary study suggests the need of the application of privacy-preserving methods for guaranteeing the privacy protection during the analysis and publishing of individual patterns. An interesting future work would involve the study of privacy methods that exploit the knowledge provided by the risk assessment methodology for reducing the model perturbations.

Acknowledgments. Work partially supported by the EU H2020 Program under the funding scheme "INFRAIA-1-2014-2015: Research Infrastructures", grant agreement 654024 *"SoBigData"* (http://www.sobigdata.eu).

References

1. Adomavicius, G., Tuzhilin, A.: Using data mining methods to build customer profiles. Computer **34**(2), 74–82 (2001)
2. Agrawal, R., Imieliński, T., Swami, A.: Mining association rules between sets of items in large databases. In: Proceedings of the 1993 ACM SIGMOD International Conference on Management of Data, SIGMOD 1993, pp. 207–216. ACM, New York (1993)
3. Agrawal, R., Srikant, R., et al.: Fast algorithms for mining association rules. In: Proceedings of the 20th International Conference on very Large Data Bases, VLDB, vol. 1215, pp. 487–499 (1994)
4. Andersen, H., Andreasen, M., Jacobsen, P.: The CRM Handbook: From Group to Multi-individual. PricewaterhouseCoopers, Norhaven (1999)
5. De Capitani Di Vimercati, S., Foresti, S., Livraga, G., Samarati, P.: Data privacy definitions and techniques. Int. J. Uncertain. Fuzziness Knowl.-Based Syst. **20**, 793–817 (2012)
6. Deng, M., Wuyts, K., Scandariato, R., Preneel, B., Joosen, W.: A privacy threat analysis framework: supporting the elicitation and fulfillment of privacy requirements. Requir. Eng. **16**(1), 3–32 (2011)
7. Dunk, A.S.: Product life cycle cost analysis: the impact of customer profiling, competitive advantage, and quality of is information. Manag. Account. Res. **15**(4), 401–414 (2004)
8. Giannotti, F., Gozzi, C., Manco, G.: Clustering transactional data. In: Elomaa, T., Mannila, H., Toivonen, H. (eds.) PKDD 2002. LNCS, vol. 2431, pp. 175–187. Springer, Heidelberg (2002). https://doi.org/10.1007/3-540-45681-3_15
9. Guidotti, R.: Personal data analytics: capturing human behavior to improve self-awareness and personal services through individual and collective knowledge (2017)

10. Guidotti, R., Coscia, M., Pedreschi, D., Pennacchioli, D.: Behavioral entropy and profitability in retail. In: IEEE International Conference on Data Science and Advanced Analytics (DSAA), pp. 1–10. IEEE (2015). 36678 2015

11. Guidotti, R., Gabrielli, L.: Recognizing residents and tourists with retail data using shopping profiles. In: Guidi, B., Ricci, L., Calafate, C., Gaggi, O., Marquez-Barja, J. (eds.) GOODTECHS 2017. LNICST, vol. 233, pp. 353–363. Springer, Cham (2018). https://doi.org/10.1007/978-3-319-76111-4_35

12. Guidotti, R., Gabrielli, L., Monreale, A., Pedreschi, D., Giannotti, F.: Discovering temporal regularities in retail customers' shopping behavior. EPJ Data Sci. 7(1), 6 (2018)

13. Guidotti, R., Monreale, A., Nanni, M., Giannotti, F., Pedreschi, D.: Clustering individual transactional data for masses of users. In: Proceedings of the 23rd ACM SIGKDD International Conference on Knowledge Discovery and Data Mining, KDD 2017, pp. 195–204. ACM, New York (2017)

14. Guidotti, R., Rossetti, G., Pappalardo, L., Giannotti, F., Pedreschi, D.: Market basket prediction using user-centric temporal annotated recurring sequences. In: 2017 IEEE International Conference on Data Mining (ICDM), pp. 895–900. IEEE (2017)

15. Guo, L., Guo, S., Wu, X.: Privacy preserving market basket data analysis. In: Kok, J.N., Koronacki, J., Lopez de Mantaras, R., Matwin, S., Mladenič, D., Skowron, A. (eds.) PKDD 2007. LNCS (LNAI), vol. 4702, pp. 103–114. Springer, Heidelberg (2007). https://doi.org/10.1007/978-3-540-74976-9_13

16. Hildebrandt, M.: Defining profiling: a new type of knowledge? In: Hildebrandt, M., Gutwirth, S. (eds.) Profiling the European Citizen, pp. 17–45. Springer, Dordrecht (2008). https://doi.org/10.1007/978-1-4020-6914-7_2

17. Pagliuca, D., Seri, G.: Some results of individual ranking method on the system of enterprise accounts annual survey. Esprit SDC Project, Deliverable MI-3 D, 2:1999 (1999)

18. Pelleg, D., Moore, A.W., et al.: X-means: extending k-means with efficient estimation of the number of clusters. In: ICML, vol. 1, pp. 727–734 (2000)

19. Pellungrini, R., Pappalardo, L., Pratesi, F., Monreale, A.: A data mining approach to assess privacy risk in human mobility data. ACM Trans. Intell. Syst. Technol. 9(3), 31:1–31:27 (2017)

20. Pellungrini, R., Pratesi, F., Pappalardo, L.: Assessing privacy risk in retail data. In: Guidotti, R., Monreale, A., Pedreschi, D., Abiteboul, S. (eds.) PAP 2017. LNCS, vol. 10708, pp. 17–22. Springer, Cham (2017). https://doi.org/10.1007/978-3-319-71970-2_3

21. Poulis, G., Loukides, G., Gkoulalas-Divanis, A., Skiadopoulos, S.: Anonymizing data with relational and transaction attributes. In: Blockeel, H., Kersting, K., Nijssen, S., Železný, F. (eds.) ECML PKDD 2013. LNCS (LNAI), vol. 8190, pp. 353–369. Springer, Heidelberg (2013). https://doi.org/10.1007/978-3-642-40994-3_23

22. Pratesi, F., Monreale, A., Trasarti, R., Giannotti, F., Pedreschi, D., Yanagihara, T.: PRISQUIT: a system for assessing privacy risk versus quality in data sharing. Technical report (2016)

23. Samarati, P., Sweeney, L.: Generalizing data to provide anonymity when disclosing information (abstract). In: Proceedings of the Seventeenth ACM SIGACT-SIGMOD-SIGART Symposium on Principles of Database Systems, PODS 1998, p. 188. ACM, New York (1998)

24. Spruill, N.: The confidentiality and analytic usefulness of masked business micro-data. In: Proceedings of the Section on Survey Research Methods, pp. 602–607 (1983)
25. Swiderski, F., Snyder, W.: Threat Modeling. O'Reilly Media, Sebastopol (2004)
26. Torra, V., Abowd, J.M., Domingo-Ferrer, J.: Using Mahalanobis distance-based record linkage for disclosure risk assessment. In: Domingo-Ferrer, J., Franconi, L. (eds.) PSD 2006. LNCS, vol. 4302, pp. 233–242. Springer, Heidelberg (2006). https://doi.org/10.1007/11930242_20
27. Trabelsi, S., Salzgeber, V., Bezzi, M., Montagnon, G.: Data disclosure risk evaluation. In: CRiSIS 2009, pp. 35–72 (2009)
28. Trasarti, R., Guidotti, R., Monreale, A., Giannotti, F.: MyWay: location prediction via mobility profiling. Inf. Syst. **64**, 350–367 (2017)
29. Tseng, V.S., Wu, C., Fournier-Viger, P., Yu, P.S.: Efficient algorithms for mining top-k high utility itemsets. IEEE Trans. Knowl. Data Eng. **28**(1), 54–67 (2016)
30. Wang, L., Li, X.: Personalized privacy protection for transactional data. In: Luo, X., Yu, J.X., Li, Z. (eds.) ADMA 2014. LNCS (LNAI), vol. 8933, pp. 253–266. Springer, Cham (2014). https://doi.org/10.1007/978-3-319-14717-8_20
31. Weng, S.-S., Liu, M.-J.: Feature-based recommendations for one-to-one marketing. Expert Syst. Appl. **26**(4), 493–508 (2004)
32. Xu, Y., Fung, B.C.M., Wang, K., Fu, A.W., Pei, J.: Publishing sensitive transactions for itemset utility. In: Proceedings of the 8th IEEE International Conference on Data Mining (ICDM 2008), Pisa, Italy, 15–19 December 2008, pp. 1109–1114 (2008)
33. Xu, Y., Wang, K., Fu, A.W., Yu, P.S.: Anonymizing transaction databases for publication. In: Proceedings of the 14th ACM SIGKDD International Conference on Knowledge Discovery and Data Mining, Las Vegas, Nevada, USA, 24–27 August 2008, pp. 767–775 (2008)
34. Yarovoy, R., Bonchi, F., Lakshmanan, L.V.S., Wang, W.H.: Anonymizing moving objects: how to hide a mob in a crowd? In: EDBT, pp. 72–83 (2009)

Exploring Students Eating Habits Through Individual Profiling and Clustering Analysis

Michela Natilli[1(✉)], Anna Monreale[1], Riccardo Guidotti[1,2], and Luca Pappalardo[2]

[1] University of Pisa, Largo B. Pontecorvo, 3, Pisa, Italy
{michela.natilli,anna.monreale,riccardo.guidotti}@unipi.it
[2] KDDLab, ISTI-CNR, Via G. Moruzzi, 1, Pisa, Italy
{riccardo.guidotti,luca.pappalardo}@isti.cnr.it

Abstract. Individual well-being strongly depends on food habits, therefore it is important to educate the general population, and especially young people, to the importance of a healthy and balanced diet. To this end, understanding the real eating habits of people becomes fundamental for a better and more effective intervention to improve the students' diet. In this paper we present two exploratory analyses based on centroid-based clustering that have the goal of understanding the food habits of university students. The first clustering analysis simply exploits the information about the students' food consumption of specific food categories, while the second exploratory analysis includes the temporal dimension in order to capture the information about *when* the students consume specific foods. The second approach enables the study of the impact of the time of consumption on the choice of the food.

Keywords: Food analytics · Individual models · Clustering analysis

1 Introduction

Nutrition is a crucial factor of an individual's lifestyle, that may influence both their physical health and subjective well-being [3,19]. On the one hand, food is a major source of pleasure, meals are an important opportunity of social aggregation in many cultures [1,22] and dining together reduce people's perceptions of inequality [13]. On the other hand, an excessive consumption, as well as a deficient intake, of specific aliments can lead to severe physical disorders [2]. In this regard, it has been showed that fast-food and sugar-sweetened drinks consumption is associated with risk of obesity and diabetes [20,23], whereas adopting a high-fiber diet can improve blood-glucose regulation [29] and consuming fruit and vegetable could have a potentially large impact in reducing many noncommunicable diseases [15]. Given the strong relationship between eating habits and individual well-being, it is important to educate the general population, and

© Springer Nature Switzerland AG 2019
C. Alzate et al. (Eds.): MIDAS 2018/PAP 2018, LNAI 11054, pp. 156–171, 2019.
https://doi.org/10.1007/978-3-030-13463-1_12

especially young people, to the importance of a healthy and balanced nutrition [5, 16].

Unfortunately, publicly available datasets describing eating habits and food consumption – such as the EFSA database[1] and the ERS database[2] – suffer from several limitations, mainly consisting in the presence of biases of diverse nature, the lack of information at individual level, the short period of data collection and the limited number of individuals involved. Being based on surveys, daily journals or food diaries in which respondents write down what they eat and drink, the information on food consumption is often incomplete and can be affected by the well-known problem of memory effect related to self-report [25], or by the bias due to the tendency in survey respondents to answer questions in a way that will be viewed favorably by others, the so-called "social desirability" bias [21]. Moreover, the great effort required by food diaries can force the survey respondents to simplify the registration of food intake, hence reducing the accuracy of the registered information [14, 18].

Motivated by the criticalness of these aspects, in this paper we propose a data-driven approach to the understanding of eating habits, leveraging the access to a real-world database describing all the meals consumed by around 82,000 students at the canteen of University of Pisa during a 7-years period. This dataset provides us an unprecedented picture of the foods consumed by young people according to their gender, age, geographic origin, course of study, both at lunch and dinner, and their possible evolution over time. We highlight that the food consumption data are collected automatically by means of students electronic cards, hence overcoming the problems that afflict survey-based data collection. The analysis of our data with data mining techniques reveals interesting patterns. In particular, we present two clustering analyses with the aim of segmenting the students under observation based on their food habits. The first clustering analysis is based on a student profile called *foodprint*, which summarizes the food consumption of each student in specific food categories. It leads to the discovery of four main groups of eating habits, which are characterized by the students' propensity to follow a healthy diet. For example, the cluster of *balanced* students describe individuals with a varied diet, including both healthy dishes and junk food, while in the cluster of *voracious* students with a more manifold and fatty diet is preferred. These results show a variation of eating habits across the population, allowing for the possibility of proper interventions to improve the students' diet and subjective well-being. The second clustering analysis exploits the information about *when* the students consume their meals. To this end, we introduce the definition of *temporal foodprints* and we apply an analytical process that uses the clustering analysis for both discovering typical food habits and identifying groups of students having similar food habits. The results of this additional clustering analysis allows us to study the impact of the seasons and the time of consumption on the food choice.

---

[1] https://www.efsa.europa.eu/it/food-consumption/comprehensive-database.

[2] https://www.ers.usda.gov/data-products/food-consumption-and-nutrient-intakes. aspx.

The paper is organized as follows. Section 2 discusses the related work. Section 3 provides a description of real food data under analysis. Section 4 describes the clustering analysis based on student food consumption, while Sect. 5 shows the clustering analysis based on the temporal food habits of students. Finally, Sect. 6 concludes the paper.

2   Related Work

The process of construction and extraction of a personal data model is generally referred to as *user profiling*. A *personal data model* contains the systematic behaviors expressing the repetition of habitual actions, i.e., personal patterns. These patterns can be expressed as simple or complex indexes [7], behavioral rules [11], set of events [10], typical actions [27], etc. On the one hand, users' profiles are employed to analyze and understand human behaviors and interactions. On the other hand, profiles are exploited by real services to make predictions, give suggestions, and group similar users [6]. Profiles can be classified as individual or collective according to the subject they refer to [12]. An *individual* profile is built considering the data of a single individual. This kind of profiling is used to discover the particular characteristics of a certain individual, to enable unique identification for the provision of personalized services. We talk about *collective* data models when personal data or individual models generated by individual profiling are aggregated without distinguishing the individuals.

In this work we propose an approach using a data modeling similar to [4,8,9], i.e., a vector-based modeling. Moreover, all these approaches and the presented one adopt clustering as methodology to extract the individual and collective patterns. In particular, in [4] the authors defined how to build individual profiles based on mobile phone calls such that the profiles are able to characterize the calling behavior of a user. By analyzing these data model three categories of users are identified: residents, commuters and visitors. Similarly, in [8] the temporal dimension of retail market data is used to discriminate between residents and tourists. A more analytical approach, similar to the one presented in this paper, is described in [9]. The authors present an individual and collective profiling of shopping customers according to their temporal preferences. However, these works focusing on the temporal dimension do not consider what the customers buy. On the other hand, in this work we also take into account the different types of food purchased per meal producing in this way more valuable profiles.

With respect to the field of food there are various work analyzing food habits, food consumption, consequences of a certain diet, etc. However, to the best of our knowledge, this project is the first attempt of using an automatic data-driven approach for extracting the groups of individuals with similar food consumption habits [17]. In the literature it has been shown various predefined groups of people having a certain relationship with certain categories of food. For example in [20,24,28] is shown how fast-food and sugar-sweetened drinks consumption is associated with risk of obesity among teenagers, the environmental influences of adolescent, and which is the result of a healthy behavior in school-aged children, respectively. With respect to adults, in [5] is examined the link between dietary habits and depression, in [23] the relationship between local food environments and obesity, and in [22] the role of food in life in various countries. In all these works the group of food consumers are predefined a priori while in this work our target is to extract group of students having similar patterns of food habits.

3   Food Dataset

As proxy of our study we have access to a dataset provided by the Tuscan Institute of Right to Study (DSU)[3] describing 10 millions of meals consumed by about 82,871 students at the canteen of University of Pisa during a period of seven years, from January 1, 2010 to December 26, 2016 (see Table 1 left for more detailed information). The cost of a meal at the canteen varies over the years and with the number of dishes composing the meal. The dataset contains also meals of 19,141 students (23%) who have free meals at the canteen.

Table 1. Basic statistics of the dataset (left) and description of meals in the Mensana dataset (right). Each meal consists of a student identifier, a timestamp and the list of dishes composing the meal.

meals:	10,034,413
students:	82,871
grant students:	19,141
free meals:	4,730,658
dishes:	950
food categories:	41
period:	2,551 days
from:	01/01/2010
to:	12/26/2016

student id	time	dishes
A4578A	18/04/2015 12:42:00	pasta with tomato sauce, chicken breast, fruit
G23T20	18/04/2015 12:43:00	mushroom risotto, salad, fruit
GE54Y7	18/04/2015 12:44:00	pasta with tomato, fruit
⋮	⋮	⋮

Each meal is described by a record indicating the student who consumed the meal, the type of the meal, and the list of dishes (e.g., pasta with tomato sauce, salad, apple, etc.) composing the meal (see Table 1 right for an example). While there are 950 different dishes in the dataset, they are grouped in 41 food categories, each containing dishes with homogeneous nutrients characteristics.

The composition of a meal in terms of the dish composing it changes according to the student's choice based on the menu available at that day. Figure 1 (left) reports that the most popular meal composition is that one with 4 items (i.e., first course, second course, side course and fruit or dessert). The second preferred composition is that one with 3 items (typically, first or second course, side dish and fruit or dessert). Only few students choose additional items with respect to the complete meal (6 items) or meals with only 1 or 2 items.

---

[3] https://www.dsu.toscana.it/.

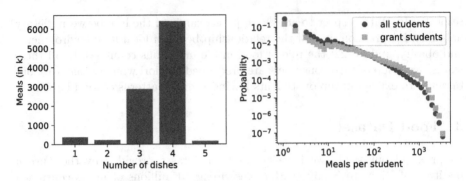

Fig. 1. Distributions: number of dishes per meal (left), meals per student (right).

Table 2. Number of students based on consumed meals

All students	
Students with at least 10 meals	60,112
Students with at least 100 meals	22,647
Students with at least 1000 meals	1,448

The total number of meals consumed at the canteen does not vary significantly over the years and it is about 1,400K each year. On the other hand, there is a slightly decrease of the number of students going to the university canteen, passing from the 30k of 2010 to the 27k of 2016. As shown in Table 2, most of the students (around 60,000) consume less than 10 meals in total, denoting the presence of a heavy-tail distribution of the number of meals consumed by the students (see Fig. 1 (right)). This means that the students going at the canteen with regularity are the minority. Only students with scholarships, for whom meals at the canteen are for free, show a slightly higher regularity.

4   Food Consumption Analytics

The dataset described in the previous section enables us to understand students' food habits by means of an appropriate vector-based user profiling. In this section we present a clustering analysis that aims at grouping students, who regularly eat at the canteen, by using information describing their typical food *consumption*.

We describe the student food consumption by using an individual model named *foodprint*. In practice, this model summarizes the consumption of the student in each food category of Table 3. We represent a *foodprint* f_u of a student u by using a vector of 41 attributes a_i with $i = 1, \ldots, 41$ (i.e., an attribute for each food category). Let $D_i^{(u)}$ be the number of dishes of the student u in the food category i and $M^{(u)}$ be the total number of meals of the student u, the value of the i-th attribute is $a_i^{(u)} = D_i^{(u)}/M^{(u)}$. We underline that if a student takes two dishes of the same category in a meal (e.g., potato dishes as second course and as side), then in $D_i^{(u)}$ is counted twice, while in $M^{(u)}$ we count only one meal.

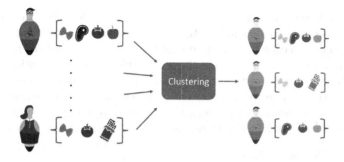

Fig. 2. Clustering process based on student foodprints

4.1 Data Preprocessing

Since we are interested in analyzing the behavior of regular students, we select a sub-population of the whole students in the dataset. In particular, we do not include in the analysis students who use the canteen sporadically by applying to the data a filter on the student meals based on the frequency in the use of the canteen. We studied the frequency distribution of the consumption of meals, and this leaded to select only students who had consumed at least 100 meals and having a distance between two meals of a maximum of 100 days. The number of days is defined considering the long periods of summer holidays or exam sessions where students may not be in Pisa. After this filtering we obtain 1,607,993 meals related to 6,890 different students. Therefore, we build a foodprint summarizing the food consumption for each one of these students.

Before proceeding with the cluster analysis we compute the correlation between the attributes of the students' profiles to identify (if present) high correlations among food categories. We observed only two strong correlations. A correlation of 0.8 between category a_{33} (soups only vegetables) and category a_{34} (potato and vegetable soups), and a correlation of 0.68 between category a_{101} (tofu o soybeans with cheese) and category a_{102} (tofu or soy with vegetables). Therefore we kept all of the categories separated using all the 41 attributes.

4.2 Clustering Analysis

The goal of the cluster analysis on the student foodprints is to identify groups of students with similar behaviors in terms of food choices. To this aim, we adopt the K-Means [26] clustering algorithm as it is shown in the literature to have good effects in grouping profiles when a vector based model is used [4,9]. K-Means requires to specify the number of clusters k as parameter. To identify the best value for k, we applied the standard approach (see [26]) that runs the clustering with several values for k and selects the one such that a further increase in k generates no significant improvement in the cluster's compactness. This aspect is measured using the Sum of Squared Error (SSE). The optimal number of clusters should be that in which the curve has a significant inflection (elbow). As consequence, we selected $k = 4$. Figure 2 describes the clustering process based on student *foodprints*, that starts with the modelling of meals and dishes of a student with a foodprint and ends by extracting students food habits.

The characterization of the four clusters provides some interesting information that allows us to understand how students behave in the university canteen.

We assigned a name to each cluster with respect to their characterization given by the food categories of the cluster centroids using the following four adjectives that well describe the feeding behavior of the students of these clusters:

- Cluster 0: *Balanced* (30.95%)
- Cluster 1: *Foodie* (17.43%)
- Cluster 2: *Health fanatic* (33.64%)
- Cluster 3: *Voracious* (17.98%)

Figure 3 depicts a graphical representation of the 4 clusters that summarizes the food profile of each cluster and that we discuss in detail in the following.

Balanced. The cluster called *balanced* covers 31% of the total observed students and most of them are males. Figure 3 depicts a radar chart showing the typical food choices of this group pf students. These students have a rather varied diet: they eat both complex dishes and healthy dishes. Moreover, they take almost always fruit and just in few occasions the dessert; they eat enough vegetables, but also pasta (or couscous) with meat.

Foodie. The *foodie* cluster contains 1,198 students and most of them are males. This cluster differs from the *voracious* cluster because these students insert a greater quantity of healthy food into their diet (see Fig. 3). In particular, 50% of their meals contain fruit while 35% contain desserts. They often consume flour with meat, cheese or eggs, but often also eat cooked or grilled vegetables and raw vegetables. Therefore, they have a rather fatty diet, but in some cases they try to include more healthy food in their diet.

Voracious. The cluster named *voracious* involves 1,235 students and most of them are females. Observing the typical food choices of this group in Fig. 3 we can see that they tend to eat mostly fatty foods. It is evident that they prefer flour with meat, they often take dessert instead of fruit and they eat a lot of dishes based on potatoes, red meat and salami. Moreover, compared to others students, voracious students have the highest consumption of fried dishes (potatoes or vegetables) and the lower consumption of cooked or grilled vegetables.

Health Fanatic. The cluster called *health fanatic* includes 2,311 students with an almost equal distribution between male (58%) and female (42%), and it is characterized by a healthy diet, as shows the radar chart in Fig. 3. Indeed, they eat very often fruit, raw vegetables and cooked or grilled vegetables. These students, compared to others, consume the lowest share of flours with meat, cheese or eggs, and they have a low consumption of red meats or salami, desserts, white

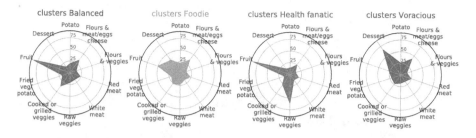

Fig. 3. Main food categories choices among clusters

meats, potato dishes and fried foods. They are the main consumers of tofu or soy-based dishes. Although the share of consume is low, we underline that the supply of these categories in the canteen is limited with respect to the others.

We also perform a study on the clusters aiming at verifying the influence of the demographical origin of students on the food choices. We observe that, looking at the origin of the students in the various clusters, the differences are not significant. There is a tendency towards the prevalence of students from southern Italy and Islands (Sicily and Sardinia) in the *foodie* group, but in general it does not seem that geographical origin is an important feature in the definition of clusters. Moreover, we also investigate the distribution of students within cluster according to their course of study, in order to verify whether there are differences among students attending humanistic courses and students attending scientific courses. Although the majority of students attends a scientific course, we find that in the *voracious* cluster there is the largest quota of students from humanistic courses.

5   Temporal Food Habits Analytics

The profiling and the clustering analysis presented in the previous section do not consider any temporal information of *when* the student consumes the meals. In this section, we extend the previous analysis that is able to group students on the basis of their food *habits* taking into consideration the time of food consumption. This clustering process, depicted in Fig. 4, is based on the definition of the *temporal foodprints* of a student, and on two tasks called *food habits discovery* and *student grouping*.

The student *temporal foodprint* is an extension of the *foodprint* f_u with the following temporal information: the *year* and the *season* of the meals consumption and the knowledge that discriminates between lunch and dinner. This extension allows us to define the typical consumption of the student in each food category during the lunch or dinner in a specific year and season (winter, spring, summer and autumn). Therefore, for each student we have the set of his temporal foodprints $F_u = \{f_u^{t_1}, \ldots, f_u^{t_n}\}$, where each $f_u^{t_i}$ is a *temporal foodprint* with $t_i = (y_i, s_i, \mu_i)$ representing the temporal information composed of three elements: y_i denoting the *year*, s_i denoting the *season*, and μ_i indicating lunch or dinner.

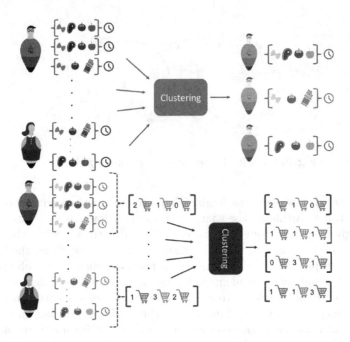

Fig. 4. Two-steps clustering process based on temporal foodprints: segmentation of the temporal foodprints *(top)* and grouping of the food habits profiles *(bottom)*

Temporal Food Habits Discovery. Similarly to the previous section, given the *temporal foodprints* of the students, we can apply a clustering algorithm to extract typical food habits by exploiting the specificity of the student consumption with respect to the temporal information t used. This clustering does not provide a students' segmentation but a segmentation of the temporal foodprints, i.e., the students temporal habits (see Fig. 4 (top)). Thus, as highlighted in Fig. 4 (bottom), a student can have his temporal foodprints distributed over different clusters, meaning that he is characterized by different food habits.

Student Grouping. The knowledge of how temporal foodprints are distributed for each student enables a student segmentation on the basis of the temporal food habits. To this aim, for each student we construct a *food habit profile* that describes the intensity of the student presence in each cluster (food habit). We represent the student habit profile through a vector of attributes h_j with $j = 1, \ldots k$, where each attribute h_j represents the intensity of a certain temporal food habit. Given the set of *temporal foodprints* of a student u, we denote by $N_{(u)}^{h_j}$ the number of his temporal foodprints belonging to the food habit h_j, and by N_u the total number of his temporal *foodprints*. Finally, we model the intensity of a student u in a cluster as $h_j = N_{(u)}^{h_j}/N_u$. We can now group the students according to their temporal food habits by re-applying a clustering algorithm on the student *food habits profiles* as shown in Fig. 4 (bottom).

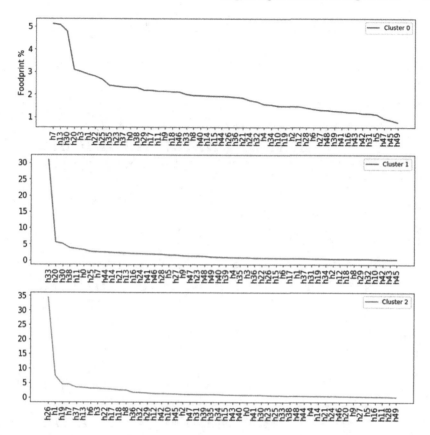

Fig. 5. Sorted levels of intensity for the temporal foodprints within the centroid of clusters 0, 1, 2. With the exception of cluster 0 which has three temporal foodprints consistently higher than the rest for all the other centroids it is possible to isolate a unique dominant temporal foodprint.

5.1   Clustering Analysis

In order to perform the double-cluster analysis described above we selected students having at least 10 meals over the whole observed period. After this filter there are 45,952 students to be analyzed. Note that, the filter is different from the previous analysis because now we are not interested only into regular students.

For discovering the temporal food habits (Fig. 4 (bottom)) we adopt again the k-means algorithm. Following the same procedure described above we select $k = 50$ as number of cluster. In other words, we fix the number of different temporal food habits to be equal to 50. The 50 different temporal habits are used to build the *food habits profiles*. For each student, his food habits profile describes how his food behavior is distributed over the different temporal food habits.

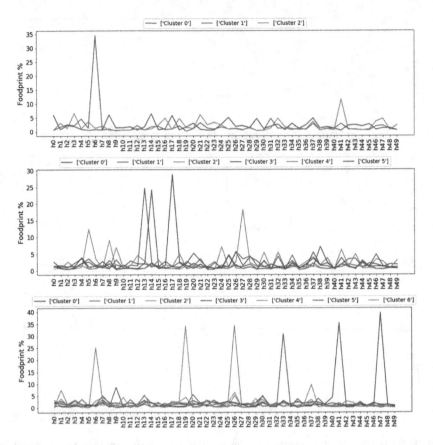

Fig. 6. Levels of intensity for the temporal foodprints using different temporal discretizations: year-lunch/dinner with three clusters, year-month-lunch/dinner with six clusters, and year-season-lunch/dinner with seven clusters, respectively.

On top of the *food habits profiles* we employ k-means for grouping the students according to their food habits. We performed this analysis using different temporal aggregations t: year and lunch/dinner information, month and lunch/dinner information, season and lunch/dinner information. In this paper we report only the results related to this last combination. However, similar results (in terms of the presence of the same pattern of having a unique dominant temporal foodprint) are obtained using the other temporal combinations. Using again the elbow method we selected $k = 7$ as number of different groups for the food habits profiles.

We analyze the clusters of the students' temporal food habits with respect to three dimensions: seasonality, time of the meals (lunch or dinner) and consumption of food categories. The first aspect we considered are the levels of intensity of the temporal food habits in the centroids of the clusters obtained. Figure 5 reports the sorted intensities for the clusters of food habits with identifiers 0, 1 and 2, respectively. On the y-axes is reported the intensity of the corresponding food habit that we have in the x-axis.

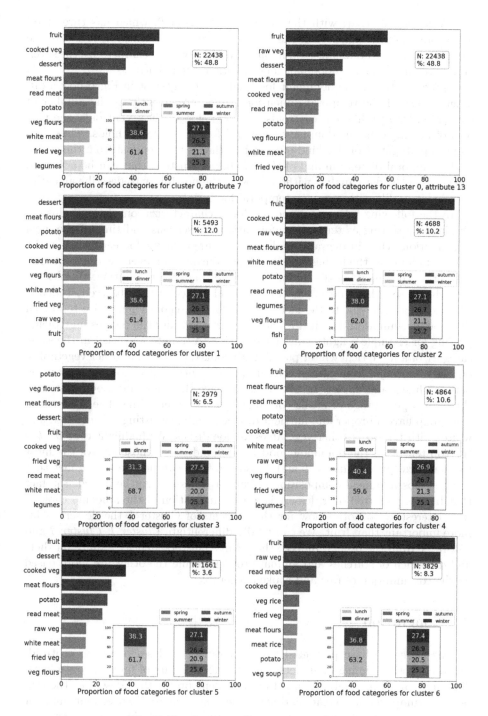

Fig. 7. Dominant attributes among clusters

We found out that, with the exception of cluster 0, which has three food habits consistently higher than the rest, for all the other centroids it is possible to isolate a unique dominant food habit. Therefore, we can characterize the seven groups by means of the dominant food habits of their centroids.

Before analyzing the dominant food habits we highlight in Fig. 6 how a different choice of the temporal dimension t and consequently of the number of clusters k, lead to different centroids. However, for all these centroids, the pattern of having a unique dominant temporal foodprint always holds. Thus, this seems a prerogative of this kind of modeling that is independent from the choice of the temporal dimension and of the number of clusters, that can be adjust to the needs of the data analyst without loosing the possibility to describe an entire food habit profile with a unique temporal foodprint.

In the following we present the results of characterization of the 7 clusters obtained by setting as temporal information the seasons and time of the meal consumption (the third graph in Fig. 6). Figure 7 shows for each cluster the information about the highest levels of food consumption over the different food categories, the impact of the season and of the time of consumption. We observe that the seasons have no impact on the food choice of the observed students. This aspect can be due to the fact that the food categories generalize too much the food consumption, i.e., it does not properly consider the recipes ingredients putting for example a light "summer pasta" with tomato and basil in the same category of a more rich "winter pasta" with cooked tomato and other vegetables. On the other hand, some differences are present with respect to the distribution between lunch and dinner. In *Cluster 4* we have the dominant food habit with the largest percentage of meals consumed during dinner time. The students in this group have a proper and nutrient dinner mainly consisting in meat (red and white) with potato or other vegetables as side dishes. We highlight that *fruit* is the top ten of all the dominant food habits, and it is almost always the most frequent, except for *Cluster 1* and *Cluster 3*. This can be due to the fact that in every possible meal composition a fruit or a dessert can be added: the fact that fruit is preferred over dessert is a good indication of the students choices. It is interesting to notice how for the students of Cluster 1 the dessert is far more important than fruit. We can finally differentiate *Cluster 1*, *Cluster 3* and *Cluster 4* from the others. Indeed, these three clusters identify a quite fat diets and diets rich of starches and animal proteins, while the other are more based on the consumption of raw and cooked vegetables.

6   Conclusion

In this paper we have presented two exploratory analyses based on centroid-based clustering aiming at understating the food habits of university students. The first clustering analysis, based on a student profile that describes the student consumption in specific food categories, allow us to discover four different clusters: *voracious*, *health fanatic*, *foddies* and *balanced*. The second analysis instead is based on a student profile that takes into account also the temporal information describing *when* the meal is consumed. This additional information together with a double clustering analysis allows us to perform a deeper analysis of the students' food habits also studying the impact of the seasons and the time of consumption on the food choice. The results of our analyses could be useful for suggest improvements to the students diet. Clearly, individual suggests might lead to privacy concerns that should be addressed appropriately. Interesting future improvements of the work include the use of food categories that introduce a lower generalization of the recipes of the consumed dishes and the link of the students' consumption with the nutrient values of the meals.

Acknowledgments. This work is part of the project Rasupea-Mensana funded by Regione Toscana on PRAF 2012–2015 funds as part of the "Nutrafood" project. Rasupea-Mensana is promoted by the University of Pisa and the Scuola Superiore Sant'Anna in collaboration with the Regional Agency for the Right to University Study of Tuscany (Azienda Regionale per il Diritto allo Studio Universitario Toscana - DSU) and Pharmanutra. This work is also partially supported by the EU H2020 Program under the funding scheme "INFRAIA-1-2014-2015: Research Infrastructures", grant agreement 654024 *"SoBigData"* (http://www.sobigdata.eu). The authors thank the staff of DSU (as part of Rasupea) and of University of Pisa for providing data and support for data linkage.

A   Appendix

Table 3. The food categories.

Cat	Foodcat	Category
a_8	Potato	Potato dishes
a_{10}	Plain flours	Plours (pasta, couscous, dumplings) in white
a_{11}	Meat flours	Flours (pasta, couscous, dumplings) with meat/cheese/eggs
a_{12}	Fish flours	Flours (pasta, couscous, dumplings) with fish
a_{13}	Veg flours	Flours (pasta, couscous, dumplings) with vegetables
a_{20}	Plain rice	Graminaceae (rice, spelled, etc.) in white
a_{21}	Meat rice	Graminaceae (rice, spelled, etc.) with meat/cheese/eggs
a_{22}	Fish rice	Graminaceae (rice, spelled, etc.) with fish
a_{23}	Veg rice	Graminaceae (rice, spelled, etc.) with vegetables
a_{31}	Meat soup	Soups with meat/cheese/egg
a_{32}	Fish soup	Soups with fish
a_{33}	Veg soup	Soups with vegetables
a_{34}	Legumes soup	Potato and legumes soups
a_{51}	Read meat	Red meat/salami
a_{52}	White meat	White meat
a_{53}	Processed meat	Meat (white/red) - processed

(continued)

Table 3. (*continued*)

Cat	Foodcat	Category
a_{60}	Fish	Fish
a_{62}	Fried fish	Fish - fried
a_{71}	Cheese salad	Cheese salad
a_{81}	Raw veg	Raw vegetables
a_{82}	Cooked veg	Cooked or grilled vegetables
a_{83}	Legumes	Legumes
a_{91}	Meat pie	Eggs, molded, pies with meat and cheeses
a_{92}	Veg pie	Eggs, molded, pies with vegetables
a_{93}	Fried veg	Vegetables/fried potatoes/other fried
a_{101}	Soy cheese	Tofu or soy with cheeses
a_{102}	Veg soy	Tofu or soy with vegetables
a_{212}	Cheese	Cheese
a_{213}	Sandwiches	Sandwiches, piadines, pizzas, bunnies
a_{415}	Fruit	Fruit
a_{416}	Dessert	Dessert

References

1. The cultural dimension of food. Technical report, Barilla Center for Food Nutrition (2017)
2. DeVault, K.R., Castell, D.O.: Updated guidelines for the diagnosis and treatment of gastroesophageal reflux disease. Am. J. Gastroenterol. **100**(1), 190 (2005)
3. Eertmans, A., Baeyens, F., Van Den Bergh, O.: Food likes and their relative importance in human eating behavior: review and preliminary suggestions for health promotion. Health Educ. Res. **16**(4), 443–456 (2001)
4. Gabrielli, L., Furletti, B., Trasarti, R., Giannotti, F., Pedreschi, D.: City users' classification with mobile phone data. In: 2015 IEEE International Conference on Big Data (Big Data), pp. 1007–1012. IEEE (2015)
5. Gillen, M.M., Markey, C.N., Markey, P.M.: An examination of dieting behaviors among adults: links with depression. Eat. Behav. **13**(2), 88–93 (2012)
6. Guidotti, R.: Personal data analytics: capturing human behavior to improve self-awareness and personal services through individual and collective knowledge (2017)
7. Guidotti, R., Coscia, M., Pedreschi, D., Pennacchioli, D.: Behavioral entropy and profitability in retail. In: IEEE International Conference on Data Science and Advanced Analytics (DSAA), pp. 1–10. IEEE (2015). 36678 2015
8. Guidotti, R., Gabrielli, L.: Recognizing residents and tourists with retail data using shopping profiles. In: Guidi, B., Ricci, L., Calafate, C., Gaggi, O., Marquez-Barja, J. (eds.) GOODTECHS 2017. LNICST, vol. 233, pp. 353–363. Springer, Cham (2018). https://doi.org/10.1007/978-3-319-76111-4_35
9. Guidotti, R., Gabrielli, L., Monreale, A., Pedreschi, D., Giannotti, F.: Discovering temporal regularities in retail customers' shopping behavior. EPJ Data Sci. **7**(1), 6 (2018)

10. Guidotti, R., Monreale, A., Nanni, M., Giannotti, F., Pedreschi, D.: Clustering individual transactional data for masses of users. In: Proceedings of the 23rd ACM SIGKDD International Conference on Knowledge Discovery and Data Mining, KDD 2017, pp. 195–204. ACM, New York (2017)

11. Guidotti, R., Rossetti, G., Pappalardo, L., Giannotti, F., Pedreschi, D.: Market basket prediction using user-centric temporal annotated recurring sequences. In: 2017 IEEE International Conference on Data Mining (ICDM), pp. 895–900. IEEE (2017)

12. Hildebrandt, M.: Defining profiling: a new type of knowledge? In: Hildebrandt, M., Gutwirth, S. (eds.) Profiling the European Citizen, pp. 17–45. Springer, Dordrecht (2008). https://doi.org/10.1007/978-1-4020-6914-7_2

13. Julier, A.P.: Eating Together: Food, Friendship, and Inequality. University of Illinois Press, Champaign (2013)

14. Livingstone, M., et al.: Accuracy of weighed dietary records in studies of diet and health. BMJ 300(6726), 708–712 (1990)

15. Lock, K., Pomerleau, J., Causer, L., Altmann, D.R., McKee, M.: The global burden of disease attributable to low consumption of fruit and vegetables: implications for the global strategy on diet. Bull. World Health Organ. 83(2), 100–108 (2005)

16. Maiorino, M.I., Bellastella, G., Giugliano, D., Esposito, K.: Can diet prevent diabetes? J. Diabetes Complicat. 31(1), 288–290 (2017)

17. Martinucci, I., et al.: Gastroesophageal reflux symptoms among italian university students: epidemiology and dietary correlates using automatically recorded transactions. BMC Gastroenterol. 18(1), 116 (2018)

18. Ortega, R.M., Pérez-Rodrigo, C., López-Sobaler, A.M.: Dietary assessment methods: dietary records. Nutricion hospitalaria 31(3), 38–45 (2015)

19. Peat, C.M., et al.: Binge eating, body mass index, and gastrointestinal symptoms. J. Psychosom. Res. 75(5), 456–461 (2013)

20. Pereira, M.A., et al.: Fast-food habits, weight gain, and insulin resistance (the CARDIA study): 15-year prospective analysis. Lancet 365(9453), 36–42 (2005)

21. Phillips, D.L., Clancy, K.J.: Some effects of "social desirability" in survey studies. Am. J. Sociol. 77(5), 921–940 (1972)

22. Rozin, P., Fischler, C., Imada, S., Sarubin, A., Wrzesniewski, A.: Attitudes to food and the role of food in life in the USA, Japan, Flemish Belgium and France: possible implications for the diet-health debate. Appetite 33(2), 163–180 (1999)

23. Spence, J.C., Cutumisu, N., Edwards, J., Raine, K.D., Smoyer-Tomic, K.: Relation between local food environments and obesity among adults. BMC Public Health 9(1), 192 (2009)

24. Story, M., Neumark-Sztainer, D., French, S.: Individual and environmental influences on adolescent eating behaviors. J. Acad. Nutr. Diet. 102(3), S40–S51 (2002)

25. Sudman, S., Bradburn, N.M.: Effects of time and memory factors on response in surveys. J. Am. Stat. Assoc. 68(344), 805–815 (1973)

26. Tan, P.N., et al.: Introduction to Data Mining. Pearson Education India, Delhi (2006)

27. Trasarti, R., Guidotti, R., Monreale, A., Giannotti, F.: MyWay: location prediction via mobility profiling. Inf. Syst. 64, 350–367 (2017)

28. Vereecken, C.A., De Henauw, S., Maes, L.: Adolescents' food habits: results of the health behaviour in school-aged children survey. Br. J. Nutr. 94(3), 423–431 (2005)

29. Zhao, L., et al.: Gut bacteria selectively promoted by dietary fibers alleviate type 2 diabetes. Science 359(6380), 1151–1156 (2018)

Author Index

Printed in the United States
By Bookmasters